THE FUSION OF ARTIFICIAL INTELLIGENCE AND SOFT COMPUTING TECHNIQUES FOR CYBERSECURITY

AAP Advances in Artificial Intelligence and Robotics

THE FUSION OF ARTIFICIAL INTELLIGENCE AND SOFT COMPUTING TECHNIQUES FOR CYBERSECURITY

Edited

M. A. Jabbar, PhD
Sanju Tiwari, PhD
Subhendu Kumar Pani, PhD
Stephen Huang, PhD

APPLE
ACADEMIC
PRESS

First edition published 2024

Apple Academic Press Inc.
1265 Goldenrod Circle, NE,
Palm Bay, FL 32905 USA

760 Laurentian Drive, Unit 19,
Burlington, ON L7N 0A4, CANADA

CRC Press
2385 NW Executive Center Drive,
Suite 320, Boca Raton FL 33431

4 Park Square, Milton Park,
Abingdon, Oxon, OX14 4RN UK

© 2024 by Apple Academic Press, Inc.

Apple Academic Press exclusively co-publishes with CRC Press, an imprint of Taylor & Francis Group, LLC

Library and Archives Canada Cataloguing in Publication

Title: The fusion of artificial intelligence and soft computing techniques for cybersecurity / edited by M.A. Jabbar, PhD, Sanju Tiwari, PhD, Subhendu Kumar Pani, PhD, Stephen Huang, PhD.
Names: Jabbar, M.A. (PhD), editor. | Tiwari, Sanju, 1979- editor. | Pani, Subhendu Kumar, 1980-editor. | Huang, Stephen (Lecturer in computer science), editor.
Description: First edition. | Series statement: AAP advances in artificial intelligence and robotics | Includes bibliographical references and index.
Identifiers: Canadiana (print) 20230587801 | Canadiana (ebook) 20230587828 | ISBN 9781774914809 (hardcover) | ISBN 9781774914816 (softcover) | ISBN 9781003428503 (ebook)
Subjects: LCSH: Computer security. | LCSH: Cyberterrorism—Prevention. | LCSH: Artificial intelligence. | LCSH: Soft computing. | LCSH: COVID-19 Pandemic, 2020-
Classification: LCC QA76.9.A25 F87 2024 | DDC 005.8—dc23

Library of Congress Cataloging-in-Publication Data

CIP data on file with US Library of Congress

ISBN: 978-1-77491-480-9 (hbk)
ISBN: 978-1-77491-481-6 (pbk)
ISBN: 978-1-00342-850-3 (ebk)

AAP ADVANCES IN ARTIFICIAL INTELLIGENCE AND ROBOTICS

The new book series AAP Advances in Artificial Intelligence & Robotics will provide detailed coverage of innovations in artificial life, computational intelligence, evolutionary computing, machine learning, robotics, and applications. The list of topics covers all the application areas of artificial intelligence and robotics such as: computational neuroscience, social intelligence, ambient intelligence, artificial life, virtual worlds and society, cognitive science and systems, computational intelligence, human-centered and human-centric computing, intelligent decision making and support, intelligent network security.

With this innovative era of simulated and artificial intelligence, much research is required in order to advance the field and also to estimate the societal and ethical concerns of the existence robotics and scientific computing. The series also aims that books in this series will be practically relevant, so that the results will be useful for managers in leadership roles related to AI and robotics, researchers, data analysts, project managers, and others. Therefore, both theoretical and managerial implications of the research need to be considered.

The book series will broadly consider the contributions from the following fields:

- Artificial Intelligence Applications in Security
- Artificial Intelligence in Bioinformatics
- Robot Structure Design and Control
- Artificial Intelligence in Biomedical and Healthcare
- Multi-Robot Intelligent Aggregation Mechanisms and Operation Platforms
- Artificial Intelligence and Learning Environments
- Advances in AI-Driven Smart System Designs
- Robot Navigation, Positioning, and Autonomous Control
- Robot Perception and Data Fusion
- Advances in Artificial Intelligence research
- Advances in AI-Driven Data Analytics and Innovation

- Application of Intelligent Systems for Solving Real-World Problems
- Hybrid Systems Design and Applications Using AI
- Robot Grabbing and Operation
- Robot Behavior Decision and Control
- Robot Motion and Path Planning
- Applications of Intelligent Systems and Computer Vision

For additional information, contact:
Book Series Editor: Subhendu Kumar Pani
Professor & Research Co-ordinator
Dept. of Computer Science and Engineering
Orissa Engineering College, Bhubaneswar, India
Email: pani.subhendu@gmail.com

AAP ADVANCES IN ARTIFICIAL INTELLIGENCE AND ROBOTICS

Advancements in Artificial Intelligence, Blockchain Technology, and IoT in Higher Education: Mitigating the Impact of COVID-19
Editors: Subhendu Kumar Pani, PhD, Kamalakanta Muduli, PhD, Sujoy Kumar Jana, PhD, Srikanth Bathula, PhD, and Golam Sarwar Khan, PhD

The Fusion of Artificial Intelligence and Soft Computing Techniques for Cybersecurity
Editors: M. Ajabbar, PhD, Sanju Tiwari, PhD, Subhendu Kumar Pani, PhD, and Stephen Huang, PhD

Incorporating AI Technology in the Service Sector: Innovations in Creating Knowledge, Improving Efficiency, and Elevating Quality of Life
Editors: Maria Jose Sousa, PhD, Subhendu Pani, PhD, Francesca dal Mas, PhD, and Sergio Sousa, PhD

Handbook of Research on Artificial Intelligence and Soft Computing Techniques in Personalized Healthcare Services
Editors: Uma N. Dulhare, PhD, A. V. Senthil Kumar, Amit Dutta, PhD, Seddik Bri, PhD, and Ibrahiem M. M. El Emary, PhD

Fusion of Artificial Intelligence and Machine Learning in Advanced Image Processing
Editors: Arun Rana, PhD, Rashmi Gupta, PhD, Sharad Sharma, PhD, Ahmad A. Elngar, PhD, , and Sachin Dhawan, PhD

Artificial Intelligence and Machine Learning for Business
Editors: Namita Mishra, PhD, and A. V. Senthil Kumar, PhD, DSc

Advances in Autonomous Navigation through Intelligent Technologies
Editors: Niharika Singh, PhD, Thipendra P. Singh, PhD, and Brian Azzopardi, PhD

ABOUT THE EDITORS

M. A. Jabbar, PhD
Professor and Head, Department CSE (AI and ML), Vardhaman College of Engineering, Hyderabad, Telangana, India

M. A. Jabbar, PhD, is a professor and Head of the Department CSE (AI and ML), Vardhaman College of Engineering, Hyderabad, Telangana, India. He obtained his PhD in the year 2015 from JNTUH, Hyderabad, and Telangana, India. He has been teaching for more than 20 years. His research interests include artificial intelligence, big data analytics, bioinformatics, cyber security, machine learning, attack graphs, and intrusion detection systems.

He has published 62 papers in various journals and conferences. He has served as a technical committee member for more than 70 international conferences. He has been the editor for 1st ICMLSC 2018, SOCPAR 2019, and ICMLSC 2020. He also has been involved in organizing international conferences as an organizing chair, program committee chair, publication chair and reviewer for SoCPaR, HIS, ISDA, IAS, WICT, NABIC, and others. He is the Guest Editor for *The Fusion of Internet of Things, AI, and Cloud Computing in Health Care: Opportunities and Challenges* (Springer) series, and *Deep Learning in Biomedical and Health Informatics: Current Applications and Possibilities*—CRC Press, Guest Editor for *Emerging Technologies and Applications for a Smart and Sustainable World*—Bentham Science, Guest Editor for *Machine Learning Methods for Signal, Image, and Speech Processing*—River Publisher.

Sanju Tiwari, PhD
Senior Researcher, Universidad Autonoma de Tamaulipas, Mexico

Sanju Tiwari, PhD, is a Senior Researcher at Universidad Autonoma de Tamaulipas (70 years old University), Mexico. She worked as a postdoctoral researcher in Ontology Engineering Group, Universidad Polytecnica De Madrid, Spain. Prior to this, she has worked as a research associate for a sponsored research project, Intelligent Real Time Situation Awareness and Decision Support System for Indian Defence, funded by DRDO,

New Delhi in Department of Computer Applications, National Institute of Technology, Kurukshetra. In this project, she has developed and evaluated a Decision Support System for Indian Defence. Her current research interests include, ontology engineering, knowledge graphs, linked data generation and publication, semantic web, reasoning with SPARQL and machine intelligence.

Subhendu Kumar Pani, PhD
Professor, Department of CSE, OEC, Bhubaneswar under BPUT, India

Subhendu Kumar Pani, PhD, is currently working as a Professor in the Department of CSE, OEC, Bhubaneswar under BPUT. He has more than 15 research articles published along with six authored books, 15 edited books and 30 book chapters to his credit. He has over 16 years of teaching and research experience. His research interests include data mining, big data analysis, web data analytics, fuzzy decision making, and computational intelligence. He is the recipient of five researcher awards. In addition to research, he has guided two PhD students and 31 MTech students. He is the (founding) series Editor of CRC Press (Taylor Francis Group) book series on *Emerging Trends in Biomedical Technologies and Health Informatics*. He is a fellow in Scientific Society of Advanced Research and Social Change and life member in IE, ISTE, SCA, OBA, OMS, SMIACSIT, SMUACEE, CSI. He also regularly serves as a program committee member for numerous national and international conferences.

Stephen Huang, PhD
Professor of Computer Science, University of Houston, Texas, USA

Stephen Huang, PhD, is a Professor of Computer Science, with more than 30 years of experience in teaching, research, and administration in Computer Science. His research interests include cybersecurity, intrusion detection, bioinformatics, data science, and algorithms. He has published many journal and conference papers, and supervised and assisted over 30 M.S. and eight PhD students. Dr. Huang has served in several administrative positions at UH including Director of Graduate Studies, Associate Chairman, and Department Chairman. Dr. Huang received his PhD degree in Computer Science from the University of Texas-Austin in 1981. He was a National Research Council-NASA Senior Research Associate at NASA Goddard Space Flight Center, Greenbelt, Maryland from 1989–1990.

CONTENTS

CONTRIBUTORS

Noora Alfurais
Department of Information Technology, Ajman University, Ajman, UAE

Yasmin Alhayek
Department of Information Technology, Ajman University, Ajman, UAE

Nujuom Assar
Department of Information Technology, Ajman University, Ajman, UAE

Ruqqaiya Begum
Vardhman College of Engineering, Hyderabad, India

R. Roopa Chandrika
Malla Reddy College of Engineering and Technology, Hyderabad, India

Neeranjan Chitare
Northumbria University, Newcastle, UK

Gaurav Choudhary
Department of Applied Mathematics and Computer Science, Technical University of Denmark (DTU), Denmark

N. Divya
Sri Ramakrishna Engineering College, Coimbatore, Tamil Nadu, India

Pavan Kumar E.
Department of ECE, Saividya Institute of Technology, Bengaluru, Karnataka, India

R. R. Rubia Gandhi
Sri Ramakrishna Engineering College, Coimbatore, Tamil Nadu, India

N. S. Gowri Ganesh
Malla Reddy College of Engineering and Technology, Hyderabad, India

Maitreyee Ghosh
Chubb UK Services Limited, London, UK

Lokesh Giripunje
School of Computer Science and Engineering (SCSE), VIT Bhopal University, Bhopal, Madhya Pradesh, India

Shankru Guggari
Machine Intelligence Research Labs, USA

M. A. Jabbar
Vardhaman College of Engineering, Hyderabad, India

Abinaya Inbamani
Sri Ramakrishna Engineering College, Coimbatore, Tamil Nadu, India

Hena Iqbal
Department of Information Technology, Ajman University, Ajman, UAE

M. Karthik
Sri Ramakrishna Engineering College, Coimbatore, Tamil Nadu, India

R Karthikeyan
Vardhaman College of Engineering, Hyderabad, India

V. Kavitha
Department of Computer Applications, Hindusthan College of Arts and Science, Coimbatore, India

M. Sivaram Kumar
Karpagam Academy of Higher Education, Coimbatore, Tamil Nadu, India

A. Mummoorthy
Malla Reddy College of Engineering and Technology, Hyderabad, India

Amisha R. Naik
Department of MCA, Dr Ambedkar Institute of Technology, Bengaluru, India

Mohan Kumar K. N.
Department of Computer Science and Engineering, Saividya Institute of Technology, Bengaluru, Karnataka, India

Piyush Kumar Pareek
Department of CSE, Nitte Meenakshi Institute of Technology, Bengaluru, India

E. Ramya
Bannari Amman Institute of Technology, Sathyamangalam, India

Bharathi S.
Department of MCA, Dr Ambedkar Institute of Technology, Bengaluru, India

Sanchari Saha
Department of Computer Science and Engineering, CMR Institute of Technology, Bengaluru, Karnataka, India

A. Dhanu Saswanth
Dept of Computer Science with Cognitive Systems Sri Ramakrishna college of Arts and Science Coimbatore, India

Smit Sawant
School of Computer Science and Engineering (SCSE), VIT Bhopal University, Bhopal, Madhya Pradesh, India

Shishir Kumar Shandilya
School of Computer Science and Engineering (SCSE), VIT Bhopal University, Bhopal, Madhya Pradesh, India

Sonia Sharma
Department of C.Sc. and Applications, Hindu Girls College, Jagadhri, Haryana, India

Vikas Sihag
Department of Cyber Security, Sardar Patel University of Police, Jodhpur, Rajasthan, India

Tanjin Sikder
Department of Information Technology, Ajman University, Ajman, UAE

B. Sundaravadivazhagan
Department of Information and Technology, University of Technology and Applied Science, Al Mussanah, Oman

Rayeesa Tasneem
Vardhaman College of Engineering, Hyderabad, India

K. Tejasvi
Vardhman College of Engineering, Hyderabad, India

ABBREVIATIONS

ABAC	attribute-based access control
ABE	attribute-based encryption
ACO	ant colony optimization
AI	artificial intelligence
AMOP	advanced message queuing protocol
ANN	artificial neural networks
APT	advanced persistent threat
BDD	binary decision diagram
BDD-ML-ELM	binary decision diagram multi-layer extreme learning machine
BEC	business email compromise
BN	Bayesian networks
BW	bandwidth
CISA	Cyber Security and Infrastructure Security Agency
CMS	content management system
CNN	convolution neural network
CoAP	Constrained Application Protocol
CSP	content security policy
DAC	discretionary access control
DCS	distributed control systems
DD	dependence diagram
DDS	Data Distribution Service
DHS	Department of Homeland Security
DDoS	distributed denial-of-service
DL	deep learning
DLP	discrete logarithm problem
DML-DIV	distributed machine learning oriented data integrity verification
DoS	denial-of-service
DP	differential privacy
DQN	deep Q-network
DT	decision tree
EC	equal coefficient

EC	evolutionary computation
ECC	elliptic curve cryptography
ETSI	European Telecommunications Standards Institute
eHIPF	environments using history-based IP filtering
FNLP	fuzzy nonlinear programming
FL	fuzzy logic
FT	fault tree
GA	genetic algorithm
GAAC	genetic ant colony algorithm
GRC	governance, risk, and compliance
HE	homomorphic encryption
HIPAA	Health Insurance Portability and Accountability Act
IaaS	infrastructure as a service
ICT	information and communication technologies
IDS	intrusion detection system
IEC	International Electro Specialized Commission
IoHT	Internet of Health Technology
IOT	Internet of Things
IPS	intrusion prevention system
ISO	International Standards Organization
KNNs	K-nearest neighbors
LM	Levenberg–Marquardt
LSTM	long short-term memory
MAC	mandatory access control
MAPE	mean absolute percentage error
MARE	mean absolute relative error
MC	Markov Chains
MDP	Markov decision mechanism
MEC	mobile edge computing
MFA	multi-factor authentication
MFCC	mail-frequency spectral coefficient
MitM	man-in-the-middle
ML	machine learning
MLP	multi-layer perceptron
MQTT	Message Queue Telemetry Transport Protocol
MRE	mean relative error
MSE	mean square error
MSPs	managed service providers

NB	Naïve Bayes
NCS	networked control systems
NCSC	National Cyber Security Center
NFV	network function virtualization
NID	network intrusion detection
NIDS	network intrusion detection system
NNs	neural networks
NRMSE	normalized RMSE
OCR	office of civil right
OWASP	Open Web Application Security Project
PaaS	platform as a service
PDP	provable data possession
PHI	protected health information
PII	personally identifiable information
PNs	Petri nets
PR	probabilistic reasoning
PRE	proxy re-encryption
RBAC	role-based access control
RBD	reliability block diagram
RCE	remote execution code
RF	random forest
RMAC	reference monitoring access control
RMSE	root mean square error
RMSEC	RMSE with cost
RS	rough sets
RSA	Rivest–Shamir–Adleman
SAN	sensor actuator networks
SaaS	software as a service
SCACO	security and cost-aware computation offloading
SCM	supply chain management
SDN	software-defined networking
SGD	stochastic gradient descent
SHFRS	soft hesitant fuzzy rough collection
SOM	self organizing maps
SPN	stochastic Petri nets
SQL	structural query language
SSO	single sign-on
SVM	support vector machine

UCOM	usage-control-based access control
VAPE	variance absolute percentage error
VM	virtual machines
VMM	virtual machine monitor
VPN	virtual private network
WFH	work from home
WHO	World Health Organization
WIA	windows integrated authentication
WISN	wireless industrial sensor networks
WSAN	wireless sensor and actuator networks
XMPP	extensible messaging and presence protocol
XSS	cross-site scripting

PREFACE

Cyber security plays a vital role and has always been indispensable for secure communication. Cyber attacks are becoming a danger to organizations, their employees, and individuals. As the world becomes more ICT-enabled, cyber threats are also on the rise. Cyber security is thus becoming increasingly challenging as well. Organizations need to keep security as a top priority and have to assure confidentiality, authenticity, and availability of customers' information. Due to increasing incidents of cyber attacks, artificial intelligence and soft computing techniques have been developed for a decision support system in cyber security. These techniques have been widely used to solve problems that are difficult to solve using conventional algorithms. The main objective of this book is to create a volume of recent works on advances in artificial intelligence and soft computing techniques in cyber security.

Section 1 comprises four chapters. Chapter 1 explains An Insight into Cyber Security During the COVID-19 Pandemic. This chapter draws special attention to the role of cyber security in COVID-19 pandemic, important cyber attacks, and impacts on healthcare and business organizations. Additionally, emerging cybersecurity technologies are discussed which are useful to reduce cyber risks and finally few mitigation steps are drawn.

Chapter 2: Cyber-Attack Surge in Covid Scenario—Reasons and Consequences explains the relationship between emotions and logical thinking in order to help avoid anxiety and to choose the most logical (safe) decision. It also introduces a brief description of cyber attacks during the COVID-19 period and focuses on effects of cyber attack on various environments.

Chapter 3 discusses the impact and challenges IoT in security during the pandemic.

Chapter 4: Mobile App Development Privacy and Security Checklist During Covid-19 is a study based on the role of the mobile phone applications in individual life and the privacy predictor of those applications.

Section 2 comprises three chapters. Chapter 5: Cloud and Edge Computing Security Using Artificial Intelligence and Soft Computing

Techniques deals with the security of edge computing and its associated environment by the application of artificial intelligence and soft computing techniques.

Chapter 6: Security in IoT Using Artificial Intelligence allows readers to be able to visualize the role of AI in correctly analyzing and preventing harmful attacks with greater efficiency.

Chapter 7 on cyber security for intelligent systems is an audit chapter that aggregates data from a few other overviews and exploration papers in regard to IoT, AI, and assaults and investigates the connection between these three subjects.

Section 3 consists of three chapters. Chapter 8: Analysis of Advance Manual Detection and Robust Prevention of Cross-Site Scripting in Web-based Services explores techniques to discover reflected cross-site scripting (XSS) attacks in URL paths and the use of XSS hunter, an online platform, to find blind cross-site scripting attacks while also boosting web applications against all types of cross-site scripting (XSS) attacks.

Chapter 9: Soft Computing Techniques for Cyber Physical Systems describes the soft computing taxonomy with special inclusion of fuzzy logic, artificial neural network, and genetic algorithm. Various integration technologies with regards to cyber physical systems are also described.

Chapter 10 explores the applications of Cyber Security Using Machine Learning Approaches.

Section 4 consists of three chapters. Chapter 11: Efficient Vehicle Tracking System in Dense Traffic to Enhance the Security discusses this kind of technique that is extensively applied in heavy traffic and high-speed vehicle-moving zones, like tolling, no parking zones, etc. The primary objective of this algorithm is designed to improve security systems.

Chapter 12 proposes Optimized Analysis of Network Forensic Attacks Using Enhanced Growing Neural Gas (GNG) Clustering Technique.

Chapter 13 talks about Security in IoT.

PART I

Cybersecurity During COVID-19 Pandemic

CHAPTER 1

AN INSIGHT INTO CYBERSECURITY DURING THE COVID-19 PANDEMIC

RAYEESA TASNEEM and M.A. JABBAR

Vardhaman College of Engineering, Hyderabad, India

ABSTRACT

As the worldwide lockdown was the only solution for controlling the coronavirus disease, social distancing caused a massive dependency on online and other alternatives of cyberspace and shifted the entire world to a digital platform. The increase in anxiety, stress, and fear among individuals because of the pandemic has given malicious hackers a chance to misuse the confidential information of users by launching cyberattacks against unsuspected victims. Though online interactions, transactions, and remote work are done effectively, alternatives of cyberspace have given rise to various cybersecurity challenges. The cyberattacks exploited the vulnerabilities of a normal user as well as the system and have greatly affected data privacy, integrity, and digital behavior. This chapter draws special attention to the role of cybersecurity in the COVID-19 pandemic, important cyberattacks, and their impacts on healthcare and business organizations. Additionally, emerging cybersecurity technologies are discussed that are useful to reduce cyber risks, and finally, a few mitigation steps are drawn.

The Fusion of Artificial Intelligence and Soft Computing Techniques for Cybersecurity.
M. A. Jabbar, Sanju Tiwari, Subhendu Kumar Pani, & Stephen Huang (Eds.)
© 2024 Apple Academic Press, Inc. Co-published with CRC Press (Taylor & Francis)

1.1 INTRODUCTION

In March 2020, the coronavirus disease had infected more than 100 countries and was officially declared a pandemic by the WHO. The entire world has now been combating this disease for more than a year. Some symptoms of this infection include fever, breathing difficulty, cough, fatigue, and loss of smell or taste.[37] The fatality rate of coronavirus disease is 2–5%.[36] Additionally, it is very obvious that it has greatly impacted the health of every individual and the economy of the entire world. The spread of virus has created a lot of anxiety, uncertainty, and a sudden change in the perspective of our lives. Most companies and organizations have started implementing WFH (Work-From-Home); people choose telemedicine due to the fear of virus infection; online schools and distance learning are implemented. The basic requirements for all these sectors include the internet, computers, cameras, firewall software, etc. Meetings are carried out via videoconferences.[1] Though the calls are encrypted for the sake of security, this pandemic has created new cybersecurity concerns. As everything is performed online, the cyberattacks are increasing in number. A few of the deadly threats to cybersecurity are spam email, ransomware, malware, malicious domains and websites, DDoS attacks, malicious messages via social media, etc.[1]

In this pandemic, healthcare systems have also become major targets for cyberattacks. Most of the patients can be recovered through the use of telemedicine, as personal visits to the doctor turned out to be very difficult in this ongoing pandemic. Access to a patient record is like a treasury for cybercriminals as it comprises the major data such as date of birth, information about the healthcare provider, insurance-related data, and genetic information—data that are impossible to change, unlike cases of credit cards being stolen.[3] These cases are specifically profitable for cybercriminals as they can sell the data of a patient multiple times, which is more than the amount a particular credit holds.[3] On the other hand, millions of fake COVID-19-related websites are designed by cyberattackers under newly registered domains for the distribution of malware.[16] Though they are removed quickly, they are again registered instantly so as to circulate misinformation and confuse users with updated news.[16]

The framework of NIST cybersecurity (CSF) includes five functions: identify, protect, detect, respond, and recover, as presented in Figure 1.1.[2] This framework is helpful for organizations and companies to overcome new cyberattacks. Though many threats are being identified by the industry of cybersecurity, it is the major challenge in this pandemic to fight, which puts people and organizations in a helpless situation globally. This chapter emphasizes the issues of cybersecurity, cyberattacks amid COVID-19, and the impacts on healthcare and business organizations.

FIGURE 1.1   The framework of NIST cybersecurity.
Source: Adapted from Ref. [2]

1.2   LITERATURE SURVEY

As digital technologies are adopted throughout the world, various facets of society have shifted online, such as shopping, industries, online business interactions, and tragically, crime. Because of its lucrative nature and even lower risk levels (cyberattacks can be launched from anywhere throughout the world), it is crystal clear that cybercrime is here to stay.[26] For the occurrence of a cybercrime, there must be three factors: motive, victim, and opportunity. The victim is targeted for an attack; motive refers to the purpose for committing an attack; and finally, opportunity refers to the chance for launching the attack. For example, an opportunity can be the vulnerability of a system or even a device that is unprotected.[26] While cyberattacks nowadays are becoming more advanced, particular victims are targeted based on the motivation of the cybercriminals, for example, for the sake of financial profits, revenge, or espionage. Opportunistic cyber-criminals always strive for the maximum profits and hence will be waiting for the perfect situations like medical issues, work pressure, mental stress, public incidents, natural disasters, and the ongoing COVID-19 pandemic, which leave the whole world disturbed and stressed. Cybercriminals take great advantage of these situations and launch various cyberattacks. In previous years, many opportunistic attacks were reported; a few of the attacks are explained below:

- Natural disasters: In the US, because of hurricane Katrina (2005), there was a huge amount of destruction in New Orleans and nearby areas. After a short time, multiple fake websites were created to appeal for donations, and many local residents received scam emails requesting to fill out personal data to get government relief funds, etc. Similar attacks and scams were observed in many natural disasters, like bush fires (2020) in Australia,[27] hurricane Harvey (2017),[29] and the Japan earthquakes.[28]
- Notable events or incidents: The heartbreaking death of Michael Jackson on June 25, 2009, left the world in a huge shock. Just after 8 h of his death, various spam emails were circulated online claiming the reason behind the death.[30] During public events or matches, for example, amid the FIFA World Cup (2018), many attempts were made by cybercriminals to launch various scams to attract people for free tickets.[31]

By considering the various types of scams mentioned above that are taking place throughout the world, it is so predictable that similar types of attacks with more advanced technologies will be launched in the current COVID-19 pandemic. The outbreak of COVID-19 has caused complete damage to the whole world. The daily routines are changed, and people have to adapt to a new reality like working from home, a lack of physical activities and social communications, and the stress of not being ready to face the pandemic. Due to these situations, many people are overwhelmed and stressed, which can also cause anxiety attacks, which increase the probability of becoming victims of cyberattacks.

According to reports, there has been a significant rise in malware attacks and scams since the pandemic started.[32] In April 2020, Google blocked 18 million phishing emails and malware that were related to COVID-19.[33] To increase the success rates of cybercriminals, they target goods in high demand like PPE kits, drugs, and testing kits for coronavirus and sell them by impersonating representatives of public organizations, for example, the WHO.[26] A group of hackers from China known as Hafnium discovered vulnerabilities in Microsoft Exchange that gave them access to the email accounts of at least 30,000 organizations in the US and 250,000 globally in 2021.[34] In response to this attack, after a month, the FBI hacked several computer systems in the US that run compromised versions of Microsoft Exchange software and eradicated malicious web shells.[34] In the month of April 2021, the REvil (Ransomware-as-a-service) gang demanded a ransom of $50 million from Apple supplier Quanta.[34] According to the estimation of IBM X-Force, REvil made a profit of at least $123 million in 2020, and 21.6 terabytes of data were stolen.[34] Just after a few months, REvil demanded a ransom of $70 million from Kaseya, striking the largest ransom to date. The attack was accomplished in two parts. Firstly, cyber-criminals exploited a zero-day vulnerability in Kaseya VSA software, due to which privileged access was provided to VSA servers, and then they deployed REvil ransomware across various MSPs (Managed Service Providers).[34]

According to reports of an IBM survey, in 2021, the average total cost of a data breach is 16.50 crore rupees. This was increased by 17.85% compared to the previous year. The cost per stolen or lost record was 5900 Rupees, increased by 6.85% over 2020.[35] Cybercriminals broke into Solar-Winds by inserting malware into the Orion software update of a vendor, which was pushed to nearly 18,000 customers in March 2019.[34] This permitted them to stay in organizations for months, and they were not even detected. Researchers of the threat now believe that cybercriminals from Russia compromised nearly 100 private corporations in the US and nine federal agencies' networks.[35] A criminal group known as Darkside breached the pipeline of colonial systems, and a major fuel supply was shut down for the east coast.[34]

In July 2021, Microsoft revealed a crucial Windows bug dubbed Print Nightmare that affects the Print Spooler of Windows and permits many users to access a connected printer.[34] After the successful exploitation of this vulnerability, cybercriminals were allowed to create new user accounts, delete or modify the existing data, and install programs.[34] This bug has an equal impact on both Windows 7 and Windows 10, and Microsoft recommends installing an out-of-band security update to avoid attacks.[34] It is evident that cybercriminals are ready to disrupt almost everything in the pandemic by targeting critical systems like healthcare and the supply chain of vaccines.[35] Hence, various guidelines and suggestions are also published to fight against these cyberattacks. These guidelines are crucial for controlling the increase in threats.[26]

1.3   ROLE OF CYBERSECURITY AMID COVID-19 PANDEMIC

As the entire world deals with the widespread coronavirus disease, a few longstanding challenges have become more prevalent. Cybersecurity is one such challenge that is more concentrated in this pandemic. This situation is an excellent opportunity for data breaches. When people are more desperate to get updates on virus spread, it becomes easier for cybercriminals to get confidential data by creating malicious links. The number of fake emails about coronavirus has increased since January 2020. One cybersecurity company in Russia discovered 500 instances of these types of scams, which were spread to 403 of its customers. Ever since the coronavirus pandemic started, the number of cyberattacks has also been increasing continuously.[1] Most of the cyber risks in the pandemic are due to the actions of people along with the failures of technology and systems.[1] The root causes of operational risks include the actions of people, for instance, intentional (theft, vandalism, and frauds), unintentional (mistakes, omissions, and errors), and inaction (knowledge, guidance, skills, and availability).[1] System failure and technology failure are caused by software (coding, setting of security, testing, configuration management, compatibility, etc.), hardware (capacity, maintenance, and performance), and failure of systems (design, complexity, integration, and specifications).[1] In light of these increasing cyber risks, greater care must be taken regarding cybersecurity by implementing the necessary steps, which are discussed in the following sections of this chapter.

1.3.1   CYBERATTACKS AMID COVID-19 PANDEMIC

During the COVID-19 pandemic period, cyberattacks can be classified into various categories, such as phishing, MitM attacks (Man-in-the-Middle), malware, ransomware, DoS (Denial of Service), and DDoS (Distributed Denial of Service) attacks. These types of attacks are briefly described below, and a few examples are shown in Table 1.1. Various advanced cyberattacks like phishing and scams related to the COVID-19 pandemic are launched by APT (Advanced Persistent Threat) groups and cybercriminals on organizations and unsuspected victims.[13,40] The whole pandemic is being exploited by them for their different motives, like malware and ransomware.[13]

- Online scams and phishing: Phishing is the most popular and effective type of attack in this pandemic and is done using various kinds of scams.[13] The success rate of this type of attack is nearly 30% and even higher sometimes. It is very easy to trigger a user to click on malicious or phishing links to open an attachment in the network of an organization. Phishing attackers achieve various goals, which include the delivery of malware, stealing of credentials and sensitive information, and financial fraud. This pandemic increased the effect of very common phishing traps. A few examples include cybercriminals making perfect use of Cyber Monday and Black Friday to exploit users, and the increase in online shopping during the pandemic made it especially effective in 2020. Therefore, there is a 600% increase in the number of phishing emails in Q1 2020.[13] Cybercriminals often use more advanced techniques to attract victims, like encryption protocols of HTTPS in their phishing links; nearly 75% of phishing links were prepared using SSL.[13] In addition, SaaS (Software as a Service) and webmail users are highly targeted phishing areas.
- Disruptive malware and ransomware: Malware attacks can take various forms, including computer system viruses, Trojan horses, worms, and spyware, and are used to achieve different goals. Different variants of malware are developed to do anything from stealing and collecting confidential data to displaying unnecessary ads to permanently damage an infected system. The most widely used malwares during the COVID-19 outbreak were cryptominers, botnets, info stealers, banking Trojans, ransomware, and mobile malwares. Ransomware is a malware that is designed to use encryption to force the target of the attack to pay a ransom demand. Once it is present on the system, the user files are encrypted, and payment demands are made for exchanging the decryption key. As the latest encryption algorithms are unbreakable with the available technology, the only possible method for recovering the encrypted files is to restore the data if backup is available or to pay the ransom

demand. In the COVID-19 pandemic, some of the ransomware variants have evolved to perform "double extortion" attacks. DoppelPaymer, Maze, REvil/Sodinokibi, Nemty, etc. are a few variants that steal the file copies prior to encryption and threaten to breach them if the organization or user refuses to pay the ransom demand. Though this trend was started with Maze in late 2019, it has been growing continuously as more groups follow and adopt it throughout 2020.

- DoS and DDoS attacks: The IT services and infrastructure, such as email, web applications, etc., of an organization are very critical to run a business. DoS attacks are developed to deny access to critical resources. This is accomplished by damaging the vulnerability in an application (causing it to crash) or by flooding a system with such large data or requests that the user will not be able to use the application (interpreting legitimate requests such that they can't be handled). Unlike a DoS attack, a DDoS attack damages multiple attack sources, spreads with the help of many hosts, and launches a coordinated DoS attack on a target that intensifies the power of the attack effectively, making the defense more complicated for users.[13,39] During the remote learning and work led by the COVID-19 epidemic, remote access solutions were major targets of DoS and DDoS attacks. During the academic year of 2020–2021, DDoS attacks on the education areas increased rapidly. In universities in the UK, the service provider of the internet, JISC, experienced a DDoS attack in the pandemic, which disrupted staff and students to access the IT services and the internet at the University.[13]

- MitM (Man-in-the-Middle) attacks: Various network protocols are safeguarded against eavesdroppers using encryption that makes the traffic impossible to read. A MitM attack circumvents these protections by breaking the connection into two pieces. An attacker creates a separate encrypted connection between a client and the server so that it can read or modify the data accordingly over the connection prior to forwarding it to the destination. These types of attacks can be defeated by using HTTPS protocols. Anyway, the increase in mobile usage makes these attacks more dangerous. Mobile apps offer less or no visibility to users about their network connections and the use of protocols that are insecure for the sake of communication and are prone to MitM attacks.

- Malicious apps: Mostly, all organizations concentrate their cyberse-curity efforts on computer systems, but mobile devices are a great threat to the cybersecurity of an organization. The increase in the usage of mobile devices by employees for their work and the access to confidential information of the company make malicious mobile applications more dangerous. These malicious mobile applications have the equal capability to exploit everything that desktop malware can do, for example, stealing sensitive information, encrypting files with ransomware, etc. In 2020, the second most common type of malware used worldwide was mobile malware. The variants of mobile malware include Necro, PreAMo, and xHelper, which are all Trojans with advanced functionalities, including click and ad fraud. Malware of mobiles generally takes advantage of vulnerabilities of operating systems; for example, the RCE (Remote Execution Code) vulnerability, which is present in the batch of 43 Android security patches in January 2021.

TABLE 1.1 Cyberattacks During the COVID-19 Pandemic.

S. No	Type of attack	Country	Security incident	Impact
1	Ransomware	Czech Republic	The Brno University Hospital is a testing laboratory of coronavirus disease experienced a cyberattack[8]	Entire IT network was shut down and surgeries were postponed
2	Ransomware	The US	In Los Angeles, the Hollywood Prebsyterian Medical Center had paid the ransom amount of $17,000 to obtain a decryption key for regaining the access of their hospital records[3]	It lost the revenue of 10 days and affected the reputation of the hospital management
3	Phishing	Germany	Phishing emails were sent to senior officials of the company that provide PPE. The phishing websites were developed to control officials by forcing them to create fake login pages of Microsoft so that they can steal credentials[1]	False information and defacement

TABLE 1.1   *(Continued)*

S. No	Type of attack	Country	Security incident	Impact
4	Phishing	Taiwan	Phishing emails included a hacking tool that can be accessed remotely. They impersonate the official of Taiwan's top contagious disease by urging users to get the coronavirus test[9]	Defacement and unnecessary panic created by recipients of email
5	DDoS	The US	The US Health and Human Services Department that heavily deals with issues related to coronavirus was the major target of DDoS attacks[10]	Interruption of responses related to COVID-19 issues
6	DDoS	France	In Paris, group of some hospitals that play a major role in fighting against the COVID-19 pandemic was targeted by DDoS attacks[10]	Disruption of access to emails and servers
7	Ransomware	The UK	The ransomware group named as Maze had published medical and personal details of several former patients belonging to medical research company of London that provides tests of COVID-19[11]	A major threat to the safety of a patient and also reputation of company
8	Ransomware	The US	The cybercrime named "Netwalker" hacked the University of California, San Francisco (UCSF), which was working on the vaccine of coronavirus disease and was forced to pay the ransom amount of $1.14[12]	Loss of revenue
9	Phishing/ Ransomware	Romania	In Romania, many hospitals had experienced ransomware attacks by sending phishing emails related to COVID-19 to damage computer systems, intervene hospital activities, and to encrypt files[7]	Exfiltration and disruption of hospital management

TABLE 1.1 *(Continued)*

S. No	Type of attack	Country	Security incident	Impact
10	Phishing	The US	COVID-19 vaccine manufacturing company Gilead Sciences, Inc. was a target of cybercriminals. Staff of this company was forced to create a fake login page of email for stealing their passwords[6]	Exfiltration and impersonation
11	Malware	The US, Europe, Middle East, and Asia	Malware named "Kwampirs" is a remote accessible Trojan that exploits vulnerabilities in a network of targeted companies in various countries[4,5]	Disruption in the supply of healthcare components

1.3.2 IMPACT OF COVID-19 ON HEALTHCARE CYBERSECURITY

The impact of the COVID-19 pandemic has phenomenally changed the functioning of healthcare sectors with the rapid increase in the implementation of remote patient monitoring and telehealth services. The threat in the field of healthcare too has become a source for ransomware, malware, phishing campaigns, breached medical records of patients, and various other cyberattacks in the field of healthcare with undesirable consequences. Most cyberattacks are successful when a healthcare company/organization uses outdated software or operating systems such as Windows XP or Windows 7 for monitoring medical devices throughout a hospital.[13] In fact, Interpol has stated that due to the COVID-19 pandemic, cybercriminals have shifted their targets from individuals and startup businesses to the government and critical health sectors. Security agencies in the US and UK were highly targeted in the healthcare sector, academic, pharmaceutical industries, and research companies, along with the task of providing uninterrupted patient care to virus-infected people.[15]

In the unprecedented COVID-19 circumstances, the Office of Civil Rights (OCR) released a notification that enforcement discretion will be exercised and penalties will not be imposed for not meeting the HIPAA (Health Insurance Portability and Accountability Act) regulations against providers of healthcare leveraging platforms of telehealth, which may not satisfy the privacy rules of HIPAA.[15] This is giving cybercriminals a favorable opportunity for the deployment of data breaches, EHR snooping, ransomware, phishing attacks, and many more.[15] Additionally, for the accommodation of a rapidly increasing number of infected people and the support of existing infrastructure of healthcare, most countries worldwide had to temporarily establish COVID-19 facilities for house-infected patients. As these facilities were created in a hurry, the priority is given to the patient's healthcare than security, and in this way, security becomes the least priority.[15] As a result, this leads to a vulnerability in the network that can be exploited easily by various malicious attacks. According to the reports of the Department of Health and Human Services, between February and May of 2021, there have been around 132 data breaches.[15] This is an almost 50% increase in reported breaches compared to the previous year.[15]

Tracking and contact tracing apps are other sources of privacy concerns. In some cases, the medical history of the patient needs to be referred for further treatment, and these patient records should be transferred to healthcare facilities created temporarily from regular hospitals using less secure technology.[15] This puts the healthcare sector at risk of "spray and pray" attacks by cybercrime groups. According to reports from Fortified, 60% of breaches in the healthcare sector in the first 6 months of 2020 were malicious attacks or IT incidents, apart from insiders.[15] Email compromises are also the most common attacks for getting access to the healthcare network and stealing patient records in the COVID-19 pandemic.[15] Fortified described that these cyberattacks are successfully executed with the help of phishing campaigns by dropping malware or ransomware, which are prevalent throughout the pandemic.

1.3.3   IMPACT OF CYBERSECURITY ON BUSINESS ORGANIZATIONS IN THE COVID-19 PANDEMIC

The pandemic of COVID-19 has introduced new challenges for business organizations as they switch to an operating model in which WFH has become the new normal these days. Business organizations are advancing their digital transformations, and cybersecurity is a major threat.[14] In spite of the fact that technology has been advancing every day, it is worth noticing that the majority of companies are still not providing a "cyber-safe" remote working environment for employees. In April 2020, according to the reports of the National Cyber Security Center (NCSC), Switzerland had experienced 350 various cyberattacks (fake websites, phishing, direct attacks on organizations, etc.), compared to the normal cases of 100–150.[14]

The major increase in cyberattacks was due to insecure working environments (e.g., internet security). Cybercriminals take the pandemic as an advantage and launch their cyberattacks by exploiting the weakness of employees who WFH and capitalizing on individual's strong interest in COVID-19-related news (malicious COVID-19 websites). Around 47% of individuals fall for phishing scams while working from home[14] (see Fig. 1.2). Another important thing to consider is that the average total cost of a data breach due to working remotely can be up to $137,000.[14] On July 8, 2020, according to the reports of the police from the city of London, more than GBP 11 million had been lost since January because of COVID-19-related scams.[14] In Switzerland, when a survey was conducted during the pandemic, one of the seven respondents had experienced a cyberattack.[14]

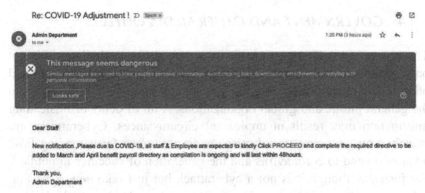

FIGURE 1.2   Phishing scams for employees at WFH.

Source: Adapted from Ref. [14]

Cybercriminals also use techniques of credential stuffing to get access to the credentials of an employee, and the stolen data are sold to other cybercriminals.[14] The major consequence of this is a serious or complete disruption to businesses which rely on platforms of videoconferencing.[14] Credential stuffing refers to the type of cyberattack where cybercriminals use stolen combinations of usernames and passwords that were stolen in the past to get access to other accounts because most individuals use the same combinations for various accounts.[14]

One of the major reasons for the increase in cyberattacks in COVID-19 is also that small, medium-sized, and startup companies do not provide computers or laptops; instead, employees use their own systems for accessing corporate data. Working remotely doesn't provide a similar level of cybersecurity compared to an office environment.[14] Employees working from home may not possibly run antimalware or antivirus scans regularly. In addition to this, home networks of Wi-Fi are easily attacked. Apart from all these attacks, the natural error committed by human is also a major area of concern. A few of the attacks are using machine learning methods, depending on the environment, and are left undetected. For example, phishing attacks have become more sophisticated via various channels like voice (vishing) and SMS. Ransomware attacks have also become more sophisticated. For example, cybercriminals are integrating data breaches with ransomware to convince targeted victims to pay the ransom amount.

1.3.4 GOVERNMENT AND OTHER MEDIA OUTLETS

The present COVID-19 pandemic is a very difficult challenge for not only healthcare and financial services but also for the government and other media resources for providing exact and immediate information to the general public and global organizations. A bit of delay or misleading information may result in unpleasant circumstances. Cyberattacks are launched on government and other media resources for spreading false updates related to coronavirus and the generation of vaccines in public.[17] For instance, though it is not a cyberattack but just fake news (see Fig. 1.3), it can also create hype and tension among people, due to which questions can be raised about governments.[17]

When the Parliament of South Africa was closed because of the restrictions of strict lockdown, all meetings were carried out using videoconferencing calls. Cybercriminals interrupted a meeting by sharing some abhorrent materials on the platform of Zoom amid the virtual meeting.[19,38] Additionally, as Zoom gained a lot of popularity amid COVID-19, it is considered unsafe in several countries and has been banned in Taiwan and the US for the sake of communication.[17]

FIGURE 1.3 Fake news related to COVID-19 vaccine.
Source: Adapted from Ref. [17]

Interpol 2020 stated that the focus of cyberattackers is on government departments and large organizations for causing unrecoverable damage to the infrastructure of information systems.[18] Due to the pandemic, the dependency on digital platforms has increased, which puts additional stress on law enforcement agencies worldwide.[18] In addition to this, most people are implementing teleconferencing tools in the space of virtual offices, and hence, hackers know the security vulnerabilities lying in these tools.[18]

1.4   COVID-19 CONSEQUENCES FOR CYBERSECURITY

The issues of cybersecurity are classified into two categories: direct and indirect consequences of the coronavirus pandemic, as shown in Table 1.2. Direct consequences include theft or fraud, data sharing, and attacks on vulnerable systems. Indirect consequences include the management of other treatments or conditions, emergency situations, and impacts on privacy and physical integrity.[20]

TABLE 1.2   Direct and Indirect Consequences of COVID-19 for Cybersecurity.

Direct consequences	Indirect consequences
Theft and fraud: Cybercriminals take advantage of people by spreading false information to create tension and stress among them so that they can easily get access to personal information (e.g., fake campaigns, ransomware, and phishing).[20] Security data breaches will be continued for a long time even after the pandemic.	Management of other treatments or conditions: Though the pandemic is a serious challenge throughout the world with an increased threat of disruption of healthcare organizations, it is impossible to delay treatments of other ongoing diseases. Patients suffering from oncological, obstetrical, chronic, mental, and various other conditions of healthcare must get treatment.[20]

TABLE 1.2   *(Continued)*

Direct consequences	Indirect consequences
Data sharing: The current issues of contact tracing apps mainly concentrate on data sharing and privacy maintenance of personal information.[20] The security vulnerabilities of these apps allow hackers to get personal data. So, contact tracing apps must be available for suitable risk evaluation and verification of data integrity.[20]	Emergency situations: Cybercriminals take advantage of people in emergency situations by sending them fraud messages or emails so that they can react to them immediately by providing personal data to cybercriminals.
Vulnerable systems: Most healthcare organizations are using outdated technology due to budget issues because of which systems become interoperable, and they also lack the latest patches and suitable configurations.[20] There's a huge risk and stress on these systems that gives a great advantage to cybercriminals to attack.	Impact on privacy and physical integrity: Advice such as consultation on web-based applications is available, but at what price for the privacy of a patient? Most home infrastructures are not integrated with security systems for controlling and protecting sensitive information.[20]

1.5 TECHNOLOGIES USED FOR FIGHTING COVID-19

The threat of cyberattacks has been increasing continuously for the last 2 years. In the past, various cyberattacks were launched, but the impact in the present situation is very severe and worse than ever before. The recent sighting of cyberattacks like NotPetya and WannaCry diverted the global focus on cybercrimes.[22] Recognition and deployment of smart cybersecurity techniques for decreasing security gaps are an absolute necessity.

- Edge computing: The focus of edge technology is on 5G or B5G (Beyond 5G), which helps to control coronavirus disease and is considered to be more powerful compared to 4G technology.[1] The edge computation integrated with the wireless network of 5G helps control the pandemic. There are various advantages of edge computing, such as low latency, protection of the training data model, and scalability.[1] A healthcare framework based on B5G technology is designed to combat COVID-19. The framework consists of three layers: cloud, edge, and stakeholder layer.[1] It is capable of integrating with the surveillance system for wearing

masks, maintaining social distancing, and testing body temperature. The designed diagnostic of COVID-19 can be useful for
identifying patients not suffering from coronavirus disease so that
hospitals are not overcrowded and sensitive personal information is
not processed.[21]

• Potential of AI (artificial intelligence) and ML (machine learning):
As AI is being introduced in every segment of the market, this
technology integrated with ML has brought massive changes in
cybersecurity.[23] AI has been paramount in designing automated
security systems and automated threat detection, face detection,
and natural language processing.[23] Technologies of AI are helpful
for protecting data against increasingly sophisticated attacks like
ransomware, malicious malware, and various social engineering
attacks. AI-enabled threat detection systems are helpful in
detecting and predicting new attacks so that administrators can be
notified of any data breach immediately. ML and AI go hand in
hand in all aspects. The integration of these emerging technologies gives rise to the rapid automation of predictive analytics.[25]
ML is useful in providing the fastest ways for the identification of
new cyberattacks, draws statistical information, and then pushes
that information to endpoint security platforms.[25] Threat intelligence is a special case where ML and AI can work tremendously
to combat cyberattacks.

• Behavioral analytics: This technique is mostly used for targeting
social networking sites and online advertisements to a suitable
audience.[22] Excitingly, behavior analytics is being explored
continuously for developing the newest cybersecurity technologies.
Behavior analytics is useful in determining patterns present on a
system and network activities for detecting potential and real-time
threats of cyberattacks.[22] For example, if the data transmission rate
increases abnormally from a particular user device, it may be an
indication of a cybersecurity issue. Additionally, this technique is
mostly useful for networks, and its application in user devices and
systems has seen rapid growth.[22]

• Blockchain cybersecurity: It is one of the current cybersecurity
technologies that is gaining recognition and momentum. The
workings of blockchain technology are based on the identification
of two transaction parties.[22] In a similar way, it also works based

on the peer-to-peer network fundamentals of this technology.[22] In a blockchain, every member is equally responsible for verifying the reliability of the data added. Additionally, blockchains develop a near-impenetrable network for cybercriminals, and the data are safeguarded from compromise. Hence, with the help of blockchain technology, AI can build a robust verification system for keeping cyberattack threats at bay.[22]

- Embedded hardware authentication: Just a PIN and a password are not enough for the protection of hardware. Embedded authenticators are the latest technologies for verifying the identity of a user. A major breakthrough was initiated by Intel in this domain by establishing sixth-generation vPro chips.[22] These user authentication chips are very powerful and are embedded into the hardware itself. They are developed to modernize "authentication security," and multiple levels and strategies of authentication are employed.[22]

- Virtual dispersive networking (VDN): This technique overcomes MitM attacks where the messages of the sender and receiver are controlled by a third party, though encryption is provided. VDN has a vital role in these types of attacks by following the rule of divide and encrypt. The messages are divided into multiple regions, and then they are individually encrypted.[24] The programming of individually encrypted messages is done on various devices like laptops, PCs, servers, and even mobile phones.[24] By using VDN, MitM attack is surpassed successfully. Cybercriminals get baffled by this technique as information is accessed in bits and pieces, which make it incomprehensible for them. VDN also increases speed and performance as soon as the traffic is dispersed independently on different paths.[24]

- Zero trust model: As the name indicates, this cybersecurity model is based on the belief that the network is already compromised.[22] By believing that the network is not trustworthy, it is obvious that both internal and external securities are to be enhanced. The main idea here is that both internal and external networks are vulnerable to compromise and need equal security.[22] It includes the identification of business-sensitive data, mapping the flow of this data, segmentation logically and physically, and then policy enforcement and control through automation and constant monitoring.[22]

1.6 MITIGATION

Mitigation and prevention of cyberattacks are quite inevitable tasks during the COVID-19 pandemic. There are a few practical ways that can be implemented for reducing the risk of cyberattacks. They are discussed below:

- Cybersecurity awareness: Awareness regarding cybersecurity must be created among users by providing constant training, which is very important to reduce the risks of cyberattacks against organizations. In a recent survey, it is clear that only 11% of businesses have trained their employees on cybersecurity for non-cybersecurity.[13]
- VPN usage: Virtual private network (VPN) is referred to as a secured encrypted channel for the transmission of data between a sender and a receiver over the internet. A VPN can be an effective barrier against cyberattacks. It offers two security aspects, integrity and confidentiality, and allows companies or organizations to extend security policies to remote employees.
- Antivirus protection: Employees of organizations must be provided with a license to malware and antivirus software so that they can use them on their personal computer systems. Also, ensure that the latest antivirus software is installed on systems and mobile devices that are running constantly.
- Phishing awareness: Employees must be very alert while receiving emails and must check the authenticity of the sender's address.[14]
- Enable MFA (Multifactor authentication): Security is strengthened by MFA as it requires a one-time code along with a username and password to get access to an account.[13] An SMS is received on a mobile device whenever anyone tries to login with a username or password. MFA is a crucial factor in reducing the risks of theft and guessing of passwords, like brute force cyberattacks.[13] An employee working from home must provide a username, password, and one-time code, which is received on his/her mobile, for the verification of his/her identity prior to accessing the company's internal network.
- Identification of weak spots: Every IT system has a weakness. Organizations should perform tests to identify them and patch the most crucial vulnerabilities without delay. The scanning of vulnerabilities in a system and various other types of penetration testing exercises must be performed on a regular basis.

- Frequent backups: Backups of important files should be performed frequently and stored independently from your system (e.g., on external drives or in the Cloud).
- Application of new tools and technology: Organizations can use the latest tools, like the host checking tool, which is used for checking the security posture of an endpoint before approving access to corporate data systems. This reinforces the security of employees working remotely.
- Physical security while working from home: It is very essential to safeguard the devices of a home office. Some of the practical ways are to ensure that work devices are not left unattended, using a lock screen for devices, setting a password for the laptop, and also making sure to log off all devices after the completion of work, etc.
- Risk management: Organizations can apply for GRC (Governance, Risk, and Compliance) solutions to enhance risk management. Solutions of GRC provide a detailed review of a company's risk exposure and help to interlink various disciplines of risk (operational risks, cybersecurity, business continuity, etc.).
- Network segmentation: Segmentation of a network is another way to reduce cyberattack risks. The network can be segmented into various trusted zones: the network of the home office (high level of trust), the network of the home entertainment (low level of trust), and the internet zone (distrusted). IoT devices must be isolated on another Wi-Fi network. So by doing this segmentation, if any of the IoT devices are compromised, it will not have any effect on a corporate laptop.
- Hospital systems should carefully monitor administrative authorities because in most cases, attacks start with compromised accounts of third party providers.[3] Regional Hospital of Hancock experienced a similar attack in January 2018.[3]

1.7 CONCLUSION

The pandemic of COVID-19 has turned the lives of every individual upside down; it is like a wakeup call for everyone. It has given many challenges to mankind, and cybersecurity is also one major challenge thrown by the pandemic that needs to be addressed very carefully to reduce cyberattacks on individuals, IT organizations, healthcare industries, etc. This chapter has reviewed the role of cybersecurity and its issues in COVID-19, outlined important cyberattacks amid the pandemic, discussed the impact of the pandemic on healthcare and business organizations, and finally noted a few mitigation steps to reduce cyber risks in the pandemic. The current situation is unchangeable, but we can try to make it better by protecting confidential information from cyberattacks.

KEYWORDS

- pandemic
- malicious
- cyberattacks
- cybersecurity
- integrity
- coronavirus

REFERENCES

1. Wang, L.; Alexander, C. A. Cyber Security during the COVID-19 Pandemic. *AIMS Electron. Electr. Eng.* **2021,** *5* (2), 146–157.
2. Weil, T.; San Murugesan, S. IT Risk and Resilience—Cybersecurity Response to COVID-19. *IT Professional* 2020, *22* (3), 4–10.
3. Williams, C. M.; Chaturvedi, R.; Chakravarthy, K. Cybersecurity Risks in a Pandemic. *J. Med. Int. Res.* **2020,** *22* (9), e23692.
4. Ying, H.; et al. Health Care Cybersecurity Challenges and Solutions Under the Climate of COVID-19: Scoping Review. *J. Med. Int. Res.* **2021,** *23* (4), e21747.
5. Cimpanu, C. FBI Re-sends Alert about Supply Chain Attacks for the Third Time in Three Months. *ZDNet,* Mar 31, 2020. https://www.zdnet.com/article/fbi-re-sends-alert-about-supply-chain-attacks-for-the-third-time-in-three-months/ (accessed Mar 2021, 05)
6. Stubbs, J.; Bing, C. Exclusive: Iran-linked Hackers Recently Targeted Coronavirus Drugmaker Gilead - Sources. *Reuters,* May 8, 2020. https://www.reuters.com/article/us-healthcare-coronavirus-gilead-iran-ex-idUSKBN22K2EV (accessed Apr 2021, 05].
7. Cimpanu, C. Hackers Preparing to Launch Ransomware Attacks Against Hospitals Arrested in Romania. *ZDNet,* May 15, 2020. https://www.zdnet.com/article/

hackers-preparing-to-launch-ransomware-attacks-against-hospitals-arrested-in-romania/ (accessed Apr 2021, 05].

8. Cimpanu, C. Czech Hospital Hit by Cyberattack while in the Midst of a COVID-19 Outbreak [Online]. https://www.zdnet.com/article/czech-hospitalhit-by-cyber-attack-while-in-the-midst-of-a-covid-19-outbreak/.

9. Lyngaas, S. 'Vendetta' Hackers are Posing as Taiwan's CDC in Data-theft Campaign [Online]. https://www.cyberscoop.com/vendetta-taiwancoronavirus- telefonica/.

10. Hale, G. DDoS Attacks on Rise due to COVID-19 [Online]. https://www.controleng.com/articles/ddos-attacks-on-rise-due-to-covid-19/.

11. Goodwin, B. Cyber Gangsters Hit UK Medical Firm Poised for Work on Coronavirus with Maze Ransomware Attack [Online]. https://www computerweekly.com/news/252480425/Cyber-gangsters-hit-UK-medical-research lorganisation-poised-for-work-on-Coronavirus.

12. Tidy, J. How Hackers Extorted $1.14m from University of California, San Francisco [Online]. https://www.bbc.com/news/technology-53214783.

13. Pranggono, B.; Arabo, A. COVID-19 Pandemic Cybersecurity Issues. *Int. Technol. Lett.* **2021,** *4* (2), e247.

14. Impact of COVID-19 on Cybersecurity [Online]. https://www2.deloitte.com/ch/en/pages/risk/articles/impact covid-cybersecurity.html

15. The Impact of COVID-19 on Healthcare Cybersecurity [Online]. https://news.sophos.com/en us/2020/10/07/the-impact-of-covid-19-on-healthcare-cybersecurity/

16. Mohsin, K. Cybersecurity in Corona Virus (Covid-19) Age, 2020 [Online]. Available at SSRN 3669810.

17. Khan, N. A.; Sarfraz, N. B.; Noor, Z. Ten Deadly Cyber Security Threats Amid COVID-19 Pandemic, 2020.

18. Chigada, J.; Madzinga, R. Cyber-Attacks and Threats during COVID-19: A Systematic Literature Review. *South Afr. J. Inf. Manag.* 2021, *23* (1), 1–11.

19. Magome, M. South Africa Sees Sharp Rise in Virus, Part of African Wave. *Associated Press*, 2020. https://www.usnews.com/news/world/ articles/2020-12-10/south-africa-sees-sharp-rise-in-virus-part-of-african-wave.

20. Ferreira, A.; Cruz-Correia, R. COVID-19 and Cybersecurity: Finally, an Opportunity to Disrupt?. *JMIRx Med.* **2021,** *2* (2), e21069.

21. Zahra, M.; Mouseli, A. Technology and its Solutions in the Era of COVID-19 Crisis: A Review of Literature. *Evid. Based Health Policy Manag. Eco.* 2020.

22. The 5 Latest Cyber Security Technologies for Your Business [Online]. https://ifflab.org/the-5-latest-cyber-security-technologies-for-your-business/

23. Top 10 Cybersecurity Tends to Watch Out For in 2022 [Online]. https://www.simplilearn.com/top-cybersecurity-trends-article.

24. Cybersecurity Technologies You Should Be A ware of in 2020 [Online]. https://dzone.com/articles/cybersecurity-technologies-you-should-be-aware-of.

25. Top 5 Technologies That Can Change The Future Of Cybersecurity [Online]. https://startuptalky.com/technologies-future-cybersecurity/

26. Lallie, H. S.; et al. Cyber Security in the Age of COVID-19: A Timeline and Analysis of Cyber-Crime and Cyber-Attacks during the Pandemic. *Comput. Secur.* **2021,** *105*, 102248.

27. Elsworthy, E. Hundreds of Bushfire Donation Scams Circulating, 2020 [Online]. https://www.abc.net.au/news/2020- 02- 07/ australia- fires- sees- spike- in- fraudster- behaviour/11923174.
28. FTC. How to Help the Earthquake Victims in Ecuador and Japan, 2016 [Online]. https://www.consumer.ftc.gov/blog/2016/04/ how- help- earthquake- victims- ecuador- and- japan (accessed June 15, 2020).
29. CNET. Watch Out For Hurricane Harvey Phishing Scams, 2017 [Online]. https:// www.cnet.com/news/ hurricane- harvey- charity- donations- scam- phishing -attack/
30. Naked Security. Michael Jackson's Death Sparks Off Spam [Online]. https:// nakedsecurity.sophos.com/2009/06/26/ michael-jackson-harvesting-email-addresses.
31. ESET. You Have NOT Won! A Look at Fake FIFA World Cup-themed Lotteries and Giveaways, 2018 [Online]. https://www.welivesecurity.com/2018/06/06/ fake- fifa- world- cup- themed- lotteries- giveaways/
32. Gallagher, S.; Brandt, A. Facing Down the Myriad Threats Tied to COVID-19, 2020 [Online]. https://news.sophos.com/en-us/2020/04/14/covidmalware
33. Kumaran, N.; Lugani, S. Protecting Businesses Against Cyber Threats during COVID-19 and Beyond, 2020 [Online]. https://cloud.google.com/blog/products/ identity-security/ protecting-against-cyber-threats-during-covid
34. Worst Cyber-Attacks of 2021 (So Far), 2021 [Online]. https://www.sdxcentral.com/ articles/news/worst-cyber-attacks-of-2021-so-far/2021/12/
35. Beware! 2022 May see Cyber Attacks Evolve to New Levels, 2022 [Online]. https:// www.thehindubusinessline.com/info-tech/beware-2022-may-see-cyber-attacks- evolve-to-new-levels/article38094515.ece
36. Zhang, J.; Wu, M. Blockchain use in IoT for Privacy-Preserving Anti-Pandemic Home Quarantine. *Electronics* **2020**, 9 (10), 1746.
37. Mohamed, Y.; Hewage, C.; Nawaf, N. IoT Technologies During and Beyond COVID-19: A Comprehensive Review. *Future Internet* **2021**, *13* (5), 105.
38. Andrade, R. O.; Ortiz-Garcés, I.; Cazares, M. In *Cybersecurity Attacks on Smart Home during Covid-19 Pandemic*, 2020 Fourth World Conference on Smart Trends in Systems, Security and Sustainability (WorldS4); IEEE, 2020.
39. Asri, S.; Pranggono, B. Impact of Distributed Denial-of-Service Attack on Advanced Metering Infrastructure. *Wirel. Pers. Commun.* **2015**, *83* (3), 2211–2223.
40. Alshamrani, A.; et al. A Survey on Advanced Persistent Threats: Techniques, Solutions, Challenges, and Research Opportunities. *IEEE Commun. Surv. Tutor.* **2019**, *21* (2), 1851–1877.

CHAPTER 2

CYBERATTACK SURGES IN THE COVID-19 SCENARIO: REASONS AND CONSEQUENCES

NEERANJAN CHITARE[1], MAITREYEE GHOSH[2],
SHANKRU GUGGARI[3], and M.A. JABBAR[4]

[1]*Northumbria University, Newcastle, UK*

[2]*Chubb UK Services Limited, London, UK*

[3]*Machine Intelligence Research Labs, USA*

[4]*Vardhaman College of Engineering, Hyderabad, India*

ABSTRACT

Cybersecurity is an essential component in reducing the effects of cyberattacks. The COVID-19 pandemic pushes for the restructuring of security strategies throughout the globe. Cybercriminals and some Advanced Persistent Threat groups use this period to attack in cyberspace. This chapter provides a relationship between emotions and logical thinking to avoid anxiety and make the most logical (safe) decision. It also introduces a brief description of cyberattacks during the COVID-19 period and focuses on the effects of cyberattacks on various environments.

The Fusion of Artificial Intelligence and Soft Computing Techniques for Cybersecurity.
M. A. Jabbar, Sanju Tiwari, Subhendu Kumar Pani, & Stephen Huang (Eds.)
© 2024 Apple Academic Press, Inc. Co-published with CRC Press (Taylor & Francis)

2.1 INTRODUCTION

The COVID-19 pandemic has created uncertainty and anxiety and enhanced drastic changes with respect to life. Most organizations adopt remote working at high speed and scale. According to the World Economic Forum, hacking and phishing are new norms. During this period, cyber-criminals increase their phishing attacks. Work-From-Home (WFH) is a new business model in this situation. Various challenges are being accomplished in 2020. Companies are building remote workforces to face the challenges by providing cloud-based platforms. 5G devices are widely used to connect various devices.[1]

During the COVID-19 pandemic period, almost all businesses adapted new operating prototypes, such as working from home. With the digital transformation, most businesses are coming across different challenges in cybersecurity. If these concerns are ignored, the organization could face severe challenges regarding operations, legality, and compliance. Almost all companies urged their employees to work from home and asked them to "stay at home" during the pandemic. This results in a crucial component for both professional and personal life. Despite the demand for technology, most companies are still not providing safe remote working environments for their staff. Before COVID-19, most business meetings were held in person, but after this pandemic, all meetings are conducted online.[2]

The organization of the chapter is as follows: Description of human errors as the source of cybersecurity breaches in Section 2; a brief explanation of the growth of the information security in Section 3; and the usage of artificial intelligence in COVID-19 is discussed in Section 4. Psychological impacts on cyberattacks are introduced in Section 5. A short note on cybersecurity issues with practical strategies to avoid cyberattacks is mentioned in Section 6. Finally, concluding remarks are presented in Section 7.

2.2 HUMAN ERROR AS THE SOURCE OF CYBERSECURITY BREACHES

Human factors in information security have remained understudied and undervalued. Human-enabled errors are to blame for the rising number of data breaches, ransomware, and cyberattacks; in fact, 95% of human-enabled errors lead to cyberattacks.[3] According to a few studies, human elements are not accounted in risk management or auditing. Technology is a key thing to handle this; most corporate executives, cybersecurity specialists, and managers depend on it. According to a few studies, some managers mistakenly believe that technology is the key to security defences.[4]

Mobile device management systems and spam filters are crucial to preserve the end-users' data. A unique risk factor, such as human error, needs to be addressed for the improvement of security. Human error is a common factor in almost all successful cyber breaches.[5] These are in various forms, such as the timely installation of security software updates, weak passwords, and disclosure of critical information in response to phishing emails.

Antimalware and threat detection softwares are developed to understand the efficiency of technical security measures and how well humans use them.[6] Generally, technical solutions halt cybercriminal activity like guessing a password to access an online company site. Human error plays a crucial role in cyberattacks. It also helps reduce a company's risk and allows you to protect your company from a much broader spectrum of risks than any single technical solution. It is very important to empower employees to deal with the new cyberattacks. Human errors are a higher priority for cyber enterprise security in 2021.[7]

2.3   THE GROWTH OF INFORMATION SECURITY BUSINESS

It is estimated that the global information security market is expected to grow at an 8.5% compound annual growth rate in the upcoming 5 years and may cross $170.4 billion during the 2022 period. It is very important to take action to understand and increase awareness about the complex challenges and technologies in this domain.[8]

Due to the growth of the digital landscape around the world, organizations need to rely on huge amounts of data. Hacktivists and cybercriminals acquire networks with the help of sharing information through the digital interface, which enables corporations to prioritize cyber defense. Furthermore, with the development of high vulnerabilities, threats, frauds, and hazards, it is very essential to design a strategy to fight against cyberattacks and help the market expand.[1]

During the COVID-19 epidemic, more cybercrime incidents were reported across different industries. More incidents happened by using malicious domain names such as COVID-19 or coronavirus. Cyberthreat actors used cybercrime as a service and targeted each demography. Similarly, fraudsters are increasingly using the same business email addresses to carry out attacks. Furthermore, the transition to a remote working model has increased the risk of cyberattacks on businesses. Organizations are urged to implement solutions for malware protection and detection. These also want to reduce growing concern about cyberthreats, which has fueled market expansion.

It is believed that the global cybersecurity market is predicted to grow at a CAGR of 10.9% during 2021–2028. It leads to cyberattacks and is responsible for the market's growth. The number of cyber frauds and crimes has increased in the last few decades, which has led to high losses for corporations. Management of the business organization is spending a huge amount on information security solutions and improving their in-house security infrastructure to combat cybercrime incidences. All over the world, most governments are giving more priority to cybersecurity plans. This is also creating high opportunities for industry players.[9]

2.3.1   INCREASE IN THE SPEAR PHISHING ATTACKS AT THE GLOBAL LEVEL

Based on the Barracuda networks, phishing emails have increased by approximately 600%. Cybercriminals try to profit from the fear and uncertainty created by the COVID-19 outbreak. Security providers reported 137, 1188, and 9116 instances in January, February, and March, respectively. COVID-19-themed emails are used for email attacks; approximately 2% is reported (i.e., 468,000 global email threats detected by the organization). The assaults took advantage of broad awareness of the matter to deceive people into passing over their logins and financial information and/or unknowingly downloading malware to their machines, as is normally the case. Here, phishing assaults are grouped as frauds (54%), blackmail (11%), brand impersonation attacks (34%), and corporate email breaches (1%). Along with the normal enticements to click through for additional information on the epidemic, some fraudsters are claiming to provide treatments or face masks, while others are attempting to obtain investment in vaccine firms or donations to combat the virus and assist the needy.[10]

In the COVID-19 epidemic, several organizations and businesses have experienced changes in their working environments. Due to the pandemic, teleworking and other distant activities have expanded. In communication, teleworking increases people's reliance on email. It is ideal for email fraud schemes. Cybercriminals have used this pandemic by exploiting broad knowledge of the issue to deceive people into giving personal information or accidentally downloading malware to their machines by clicking on harmful links or attachments. To disguise themselves as reputable sources, they may imitate government organizations, public health centers, ministries of health, or prominent figures in a relevant country. The emails appear to be genuine, as does the inclusion of the logos.[11]

2.3.2 MONETARY VALUE AND THE CYBER RISK

According to Accenture research on "Securing the Digital Economy," businesses are never more reliant on the digital economy and internet for growth. Before 10 years, one out of four relied on the internet for commercial aspects, but nowadays, without it nothing is possible. A trustworthy digital economy is very essential as well as crucial for the organization's future growth. Due to the cyberattacks, digital innovation is facing new hazards. According to the source, the United States of America expects $5.2 trillion risks in upcoming 5 years (Cumulative 2019–2023). Very huge amount of money will be lost if the investments in security are not made properly. The risk varies from industry to industry such as $753 billion for Tech, $642 billion for life sciences, and $505 billion for automotive.[12]

Due to internet dependency, the digital economy is raised. Business leaders (68%) believe that cybersecurity threats are grown tremendously. Currently, 80% of businesses are using digitally powered innovation and it is very essential to protect them from cyberattacks. According to WEF Global Risks Report 2019, cyberattacks and data fraud or theft are two venues of hazards that CEOs are most likely to challenge.[13]

In today's world, cybercrime is no more entirely the domain of rogue hackers working alone; instead, cybercriminals are forming institutionalized and organized groups. Cyber methods, crimes, and tricks make it difficult for businesses to become immune to it. New approaches for masking attacks have also been discovered, particularly with the usage of encryption protocols to conceal exploits. Malicious code identification is more challenging. Cybercriminals are increasingly employing Bitcoin to conduct transactions in order to conceal their identities. That financial gain is the driving force behind cybercrime, demonstrating how valuable this industry is. As a result, it is growing at such a quick rate. The entry fee is quite low and inexpensive. It has been noticed that cyber thieves seek a higher return on investment by employing more cost-cutting approaches, which is why complicated harmful programs, such as rootkits and bootkits, are rarely employed due to their high cost. This large investment may yield a lower return if the desired outcome is not achieved. As a result, organized hacker groups are focusing on mass-produced software that is specifically designed for cybercrime. Small businesses that have large businesses as clients are at a higher risk of being attacked because large businesses have easy access to a lot of information. Due to budget constraints, small businesses invest less in security. This is why hackers are fully aware that while their actual target may have strengthened their security, the vendors and service providers who assist them may not have. They use this as an opportunity to lapse as their entry point.[14]

During the COVID-19 period, most popular IT companies adopted various strategies for their daily operations, such as disaster recovery strategies, personal management, etc. This period also allowed understanding the weaknesses in the planning and implementation of the IT systems.[19]

2.4 AI AS A WEAPON FOR CYBERATTACK

Machine learning has previously been used to identify and fix security flaws, but can it go much further? Humans have been the exclusive authors of malicious software up until now. However, malicious software capable of changing, concealing, replicating, and reasoning on its own has proved challenging to anticipate. Humans, on the other hand, can now create programs that can "think" autonomously and attack hundreds of targets at once. The unthinkable has become unavoidable. Now that we know that expert systems are technically possible, we must prepare for a future software program capable of analyzing all possible attack vectors, selecting the best one, successfully executing it, and remaining undiscovered. These campaigns could potentially be targeted at your company. An algorithm-driven cyberattack program will operate nonstop, 24 hours a day, 7 days a week, critically "thinking" about how to attack. Most crucially, it can adapt to avoid detection. For example, these systems could be sophisticated enough to figure out every single individual who works or has ever worked for your organization, perhaps by sifting through LinkedIn data. It may then launch an assault on each of their home networks, waiting for one to connect to the corporate network. The planted malware just rides into the network and extends its wings to reach its target data, regardless of what authentication, virtual private networks (VPNs), or firewalls are in place. Disrupt, steal, or destroy; whatever the goal, it has a chance of succeeding.[15]

A few of the largest global and most reputable companies have already been harmed by cyberattacks, jeopardizing their capacity to protect sensitive data. With aggressive AI on the horizon, businesses must develop new defenses to combat it. In collaboration with AI cybersecurity firm Darktrace, MIT Technology Review Insights polled more than 300 C-level executives, directors, and managers around the world to learn how they're dealing with cyberthreats and how AI can help them combat them.

It is a known fact that human-driven reactions to cyberattacks are now not able to address the automated attacks, according to 60% of respondents, and as firms prepare for a greater challenge, more advanced technologies are necessary. In fact, a whopping 96% of respondents said they've already started guarding against AI-powered attacks, with some even enabling AI defenses. The FBI warned in January 2020 that deepfake technology had already advanced to the point where artificial identities might pass biometric tests. National security might be jeopardized by high-definition, phony videos constructed to impersonate public personalities so that they appear to be uttering whatever words the video designers put in their modified mouths, according to an FBI official at the time.

As they battle to defend data and other assets, cyberattacks are already proving to be too quick and furious for humans, and first-generation tools to keep up with traditional security tools were exposed once again in December 2020, when an operation attributed to Russian intelligence groups invaded the software supply chains of several of the world's most famous institutions, including sections of the United States government and Fortune 500 firms. Hackers recently attempted to interrupt the delivery of coronavirus vaccines, putting public health and safety at risk.

In February 2021, hackers broke into the systems of a water treatment facility in Oldsmar, Florida, to raise the quantities of pollutants in the water supply to dangerous levels. Companies are concerned that they do not have enough resources to combat such dangers.

The growing digital complexity exacerbates the IT skills gap; 60% of respondents are unable to keep up with automated attacks. AI is increasingly being used by cybersecurity teams to prevent risks from escalating at the first symptoms of compromise, allowing them to contain attacks even when they occur late at night or on weekends. When asked how concerned they are that future cyberattacks on their companies will use AI, 97% of respondents said future AI-enhanced attacks are alarming, with 58% saying such cyberattacks are extremely concerning. When asked which assaults are most concerning, the majority of respondents (68%) said impersonation and spear phishing are the most concerning. Humans are unable to keep up with the rate of AI invention, let alone respond quickly enough, necessitating new technical solutions. Thousands of businesses rely on artificial intelligence (AI) to respond to a rapidly evolving cybersecurity crisis, whether or not their security teams are present.[16] Some popular available cyber intrusion detection datasets that are available in the literature are indicated in Figure 2.1.

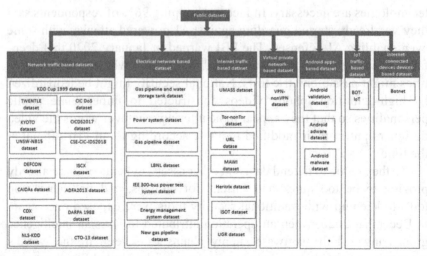

FIGURE 2.1   Public datasets for cyber intrusion detection.

2.5 PSYCHOLOGICAL EFFECTS IN COVID-19 PERIOD

During this period, most people searched for information regarding the guidelines to maintain their emotions and behaviors in this situation. At the individual level, it has impacts like social isolation enabling introspective reflection and improving one's evaluation of oneself and society. Similarly, there is an increase in predictive thinking to resolve problems and develop strategies to adopt a new balance in life at the cognitive level. This period also forced people to choose new habits for their daily lives. Curiosity among the people increases during this time if there is a failure in communication regarding official media updates, and they try to access the information from all possible authentic sources.[17]

2.6 CYBERSECURITY ISSUES IN THE ERA OF COVID-19

Cyberattacks are categorized into scams, phishing, and Distributed Denial-of-Service (DDoS). Ransomware, DDoS, and phishing are some examples of cyberattacks. Cybercriminals use this time for more commercial gain or to collect information regarding COVID-19 vaccines by adopting various approaches. Hades, patchwork, etc. are the examples adopted by the APT.

The healthcare sector is the most targeted domain in this period. Healthcare bodies, research organizations, and pharmaceutical companies are highly vulnerable to cyberattacks. Limited budgets and the usage of outdated softwares are the main reasons for these attacks. Various agencies around the globe, such as the Cybersecurity and Infrastructure Security Agency (CISA), the United Kingdom's National Cyber Security Center (NCSC), and the United States Department of Homeland Security (DHS), are working to understand cyberattacks. "WellMess" and "WellMail" are well-known custom malware used to cause damage.[18]

Practical techniques to avoid cyberattacks are as follows:

1. Need to provide constant training regarding the cybersecurity to minimize the risks.
2. Usage of VPN to communicate between two points. This provides high security in terms of integrity and confidentiality.
3. Multifactor authentication, such as username and password, and a one-time security code sent to the registered mobile phone through the message.

4. Keep all devices on the latest firmware.
5. Provide the latest antimalware software to reduce cyberattacks.
6. Divide the networks into various trusted zones.

Figure 2.2 shows a few examples of cyberattacks during the COVID-19 pandemic period.

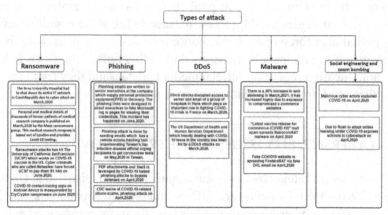

FIGURE 2.2   Cyberattacks during the COVID-19 period.

2.7 CONCLUSIONS

An increase in curiosity among the people gives cybercriminals the opportunity to develop more attacks related to the COVID-19 pandemic, such as treatment, vaccination, or methods to protect them. In this chapter, issues related to cyberattacks and how technology can help restrict these attacks are discussed. It also gives brief information regarding psychological aspects among the people.

KEYWORDS

- cybersecurity
- cyberattacks
- advanced persistent threat
- emotions, logical thinking

REFERENCES

1. Sobers, R. 134 Cybersecurity Statistics and Trends for 2021 | Varonis. *Varonis*, Mar. 16, 2021 [Online]. https://www.varonis.com/blog/cybersecurity-statistics/ (accessed Sep 05, 2021).
2. Nabe, C. Impact of COVID-19 on Cybersecurity. *Deloitte*, 2021 [Online]. https://www2.deloitte.com/ch/en/pages/risk/articles/impact-covid-cybersecurity.html (accessed Sep 05, 2021).
3. Devon, "15 Alarming Cyber Security Facts and Stats | Cybint," *Cybint*, Dec. 23, 2020 [Online]. https://www.cybintsolutions.com/cyber-security-facts-stats/ (accessed Sep 05, 2021).
4. Nobles, C. Botching Human Factors in Cybersecurity in Business Organizations. *HOLISTICA – J. Bus. Public Adm.* **2018,** *9* (3), 71–88. DOI: 10.2478/hjbpa-2018-0024.
5. Chanti, S.; Chithralekha, T. Classification of Anti-phishing Solutions. *SN Comput. Sci.* 2020, *1* (1). DOI: 10.1007/s42979-019-0011-2.
6. Neupane, A.; Saxena, N.; Maximo, J. O.; Kana, R. In: *Neural Markers of Cybersecurity: An fMRI Study of Phishing and Malware Warnings*, IEEE Transactions on Information Forensics and Security, 2016; 11 (9), 1970–1983. DOI: 10.1109/TIFS.2016.2566265.
7. Why Human Error is #1 Cyber Security Threat to Businesses in 2021, *The Hacker News*, 2021 [Online]. https://thehackernews.com/2021/02/why-human-error-is-1-cyber-security.html (accessed Sep 05, 2021).
8. Contu, R. Forecast Analysis: Information Security, Worldwide, 2Q18 Update. *Gartner*, 2018 [Online]. https://www.gartner.com/en/documents/3889055 (accessed Sep 05, 2021).
9. Cyber Security Market Trends & Growth Report, 2021-2028. *Grand View Research*, 2021 [Online]. https://www.grandviewresearch.com/industry-analysis/cyber-security-market (accessed Sep 05, 2021).
10. Muncaster, P. COVID19 Fears Drive Phishing Emails Up 667% in Under a Month - Infosecurity Magazine. *Inforsecurity*, 2020 [Online]. https://www.infosecurity-magazine.com/news/covid19-drive-phishing-emails-667/ (accessed Sep 05, 2021).
11. Understanding and Dealing with Phishing During the COVID-19 Pandemic — ENISA. *European Union Agency for Cyber Security*, 2020 [Online]. https://www.enisa.europa.eu/news/enisa-news/understanding-and-dealing-with-phishing-during-the-covid-19-pandemic (accessed Sep 05, 2021).
12. Abbosh, O.; Bissell, K. Securing the Digital Economy: Reinventing the Internet | Accenture. *Accenture*, 2021 [Online]. https://www.accenture.com/us-en/insights/cybersecurity/reinventing-the-internet-digital-economy (accessed Sep 05, 2021).
13. Global Risks 2019 : Insight Report by World Economic Forum, 2019 [Online]. http://wef.ch/risks2019 (accessed Sep 05, 2021)
14. Dalyalan, M. Cyber Risks, the Growing Threat. *Researchgate*, Nov. 2017 [Online]. https://www.researchgate.net/publication/320753018_Cyber_Risks_the_Growing_Threat (accessed Sep 05, 2021).

15. Miley, K. AI-powered Cyber Attacks. *F5 Labs- Application Threat Intelligence,* 2020 [Online]. https://www.f5.com/labs/articles/cisotociso/ai-powered-cyber-attacks (accessed Sep 05, 2021).

16. Preparing for AI-Enabled Cyberattacks. MIT Technology Review, 2021 [Online]. https://www.technologyreview.com/2021/04/08/1021696/preparing-for-ai-enabled-cyberattacks/ (accessed Sep. 05, 2021).

17. Andrade, R. O.; Ortiz-Garcés, I.; Cazares, M. In *Cybersecurity Attacks on Smart Home During Covid-19 Pandemic,* 2020 Fourth World Conference on Smart Trends in Systems, Security and Sustainability (WorldS4), 2020; pp 398–404.

18. Pranggono, B.; Arabo, A. COVID-19 Pandemic Cybersecurity Issues. *Int. Technol. Lett.* **2021,** *4,* e247.

19. Weil, T.; Murugesan, S. IT Risk and Resilience—Cybersecurity Response to COVID-19. IT Professional **2020,** *22* (3), 4–10.

CHAPTER 3

IOT IN SECURITY: IMPACT AND CHALLENGES DURING THE PANDEMIC

SONIA SHARMA

Department of C.Sc. & Applications, Hindu Girls College, Jagadhri, Haryana, India

ABSTRACT

This chapter explores the impact and challenges of security in implementing the IoT. For this chapter, we led a study on the new proposed IoT gadgets expecting to help medical care laborers and specialists during the COVID-19 pandemic. We looked into the IOT-related innovations and their executions in three stages, including "Early Diagnosis," "Quarantine Time," and "After Recovery." IoT innovation can be very effective for this pandemic; however, it is likewise basic to think about the security of information. This chapter will suggest solutions to face various challenges in securing the IoT.

3.1 INTRODUCTION

We have to put in considerable effort to get information needed from the world that we are living in. A number of examples may come to our mind. We need to send a packet to someone we knew and for this we might access a diary where the addresses are written. For this we need to locate our diary or carry it with us. When we are in a new town or city and we

The Fusion of Artificial Intelligence and Soft Computing Techniques for Cybersecurity.
M. A. Jabbar, Sanju Tiwari, Subhendu Kumar Pani, & Stephen Huang (Eds.)

need to find out a good doctor or a hospital and how do we find it out? Probably we will ask someone or look at telephone yellow pages. Often when we ask someone we might not get the correct information and we then may have to ask many others either to get complete information or even to get confirmation of the needed information. Similarly, we would like to reach a place and we will ask one or even many persons for guidance and then we receive a number of instructions on how to reach the place such as—go straight for 200 m, then take a left turn, go on straight till you find petro pump, then take the next right turn. It is difficult to remember all this information and often we are not sure about the instructions when the road has a Y junction, etc. We wish that we could get our information in a much smarter way. Yes, we have some of those smart devices already around us like our mobile phone with all the telephone numbers and addresses that we have stored. Then we have to carry our mobile phone with us. What would happen if we need information that is not stored on the mobile phone? Yes, we have a way—go to computer and search on the web. Is that easy? We need to access the Internet (luckily we can access it from our mobile, but then we have to carry our mobile!) and information gathering on the net is not very easy—our web usage is not smart enough. What we need is a smarter world that supports us with all our needs as we think of them or express them and preferably does not expect us not to carry anything and instead all the devices needed to help us are in the environment. That is our dream of our brave new world, indeed a smart world. Let us look at this world a little bit more closely. Let us take an example of our home. When I reach home, the home identifies me and opens the door, switches on the lights or opens the window curtains. At the right time (when I normally watch TV), it switches on the TV and the desired channel. It switches off the lights when I leave a room, warms the room or cools the room depending upon the time of the year/environment temperature, places order to a shop for the exhausted items in the store/refrigerator, etc. When I wanted to access the computer or web, it provides the access through the TV monitor or through mobile so that I can browse or do computation. When I leave the house it switches off the lights, locks the doors. This is a smart home. What do we need to build such a home? The door probably has a mini camera and a tiny computational device with some memory and application embedded in it. It senses the visitor at the door and runs the application. The application matches the image of the visitor with that of the stored data and would take appropriate action. The house has sensors embedded in the rooms, which detect the presence of person and identify the person and will take actions that are preferred by that person. Typically in the above description the persons concerned carry no other equipment or sensors. The sensors and processors are embedded into the environment and these devices monitor the environment or person continuously and take appropriate actions.

We can think of other example applications:

Application 1: A person with a cardiac problem wears an under garment that has a set of sensors, mote (a processor and communication device), and a tiny speaker. The sensors continuously monitor the person and the processor looks for the signals for heart attack, and when the event occurs, the processor gets in touch with the nearest hospital and gives information about the case, provides all the information of the patient records, warns the patient of the problem, advises him to go to the specified hospital, informs the nearby persons to help the patient to reach the specified hospital, tells the persons of what other actions to be taken, etc. Here the person wears his sensors, processor, and communication devices, they draw energy from body heat, the city has the infrastructure to receive messages from the patient, it has embedded processors to search for the desired hospital, inform hospital, retrieve patient records that are available on the net somewhere (in the cloud), provide access of the records to the hospital, advise the patient, etc.

Application 2: A person visits a museum, the person's profile (preferred language, interests) is accessed by the museum, visitor is provided with a WiFi hearing device and when the person walks near any exhibit, he can hear the explanation about the exhibit (with the amount of explanation suiting the interest of the visitor), etc.

Application 3: When a disaster like train accident occurs, the sensors on the train determine the nature of the accident and location, inform the railways of the accident, inform the nearest police station of the accident and advise what action is to be taken, inform the nearest hospitals, and prepare them to receive the victims. Here again the sensors, processors, and the applications are deployed in the environment, and take the desired actions when needed. Sometimes, human beings in and around the train also could inform one or more of the above groups.

The core technological ingredients of the above-mentioned systems include low-cost sensors (hard sensors) embedded in the environment and human sensors acting through different social networks or mobile networks (soft sensors) that enable pervasive sensing and information or data gathering, decision making, and taking appropriate actions in a distributed manner. The characteristics of the smart infrastructure are that the data is obtained by various sensors, owned by different interest groups and different applications, use data in different formats, use different technologies for collection and interpretation of data, use different ontologies, and use solutions in different contexts. It is a dream that exists partly today and is also distant. It offers a new world that is full of promises and pitfalls.

We can think of many more applications such as medical care from remote location, protection of property or space from intruders, and smart supermarkets which access our list of purchases and our history and guide us through purchases by telling us shelf details or describe the product, or even suggest alternative products. In all above applications, a smart world is presenting where each and every device is able to communicate and understand human interest. The basis of all this is a new computing paradigm, which names as Internet of things (IOT) where linking of objects is possible without interference of human being[1] and the devices which are used for investigation and exchange of data can be termed IOT devices.[2] The concept IOT first given by Kevin Ashton while implementing RFID concept in SCM (Supply Chain Management)[3] has given a novel research idea in the field of Internet of Things (IOT). It has acquired persuading research ground as another exploration theme in almost all the areas such as smart home, smart space, but it plays a very important role particularly in medical services. It not only reshapes the present medical field via fusing mechanical, monetary, and social possibilities, but also it develops medical care frameworks in a different way for analyzing, checking, and proper treatment of patients from traditional to more customized form. IOT is progressively turning into a crucial innovation in medical care frameworks where it can convey lower expenses, a superior nature of administrations, and progressed client encounters.[4-7] Because of its wide abilities including following, distinguishing proof and verification, and information assortment, the outstanding development of IOT in medical services is relied upon to ascend from USD 72 billion out of 2020 to USD 188 billion out of 2025.[1,8]

The current worldwide test of the pandemic brought about by the original extreme respiratory disorder COVID-2 presents the best worldwide general wellbeing emergency since the pandemic flu episode of 1918.[9] As indicated by the last report of the World Health Organization (WHO), as of September 2020, the quantity of affirmed COVID-19 cases passed 31 million individuals with a rough enormous loss of life of 960,000 individuals.[10] This sickness has comparable manifestations as seasonal influenza like fever, hack, and weariness, which are fundamental to perceive for early determination.[11] The hatching time of COVID-19 takes from 1 to 14 days. Shockingly, a patient with no indications might potentially be a transmitter of the COVID-19 infection to other people. This is while isolating such individuals is essential.[12] In addition, the recuperation time of this sickness differs and relies upon the patient's age, basic circumstances, and so forth, yet overall it can take between 6 and 41 days.[13] While this sickness has a high potential to be spread effectively in correlation with comparable illnesses inside the COVID family, there are numerous continuous endeavors and much exploration to moderate the spread of this infection. In this specific circumstance, IOT innovation has been demonstrated to be a protected and effective approach to managing the COVID-19 pandemic.[14,15]

3.2 IMPACT OF IOT DURING COVID

In all over the world the actual importance of IOT was observed by the scientist during the three major stages of COVID-19 such as early detection, quarantine period, and recovery period[17] because at that time infection of COVID-19 was spreading at increasing rate and there was necessity of such devices which can diagnose the patients in quick manner. As soon as the analysis of patients done with IOT devices, patients get better treatment and controlling of infection can be possible. Various IOT devices not only helped to detect the patients but also collect the required information by taking temperature of body during early stage of COVID-19 and monitoring of patients remotely in the second phase, that is, quarantine period.[18] These devices are capable of cleaning the areas without interference of human being and can note the activity of patients. The devices which are approved by scientists during COVID include wearable devices drones, robots IOT buttons, and various smartphone applications.

During pandemic the development of wearable devices[18] has a wonderful brunt for the detection of COVID and use of wearable devices is considered as best way to diagnose the infection in early stage. There are so many wearable devices such as smart thermometer, smart helmets, glasses, Easy-Band, and Q-Band which accept input and process it; it is either worn or attached to the body of the patient. Smart thermometers have the capability to monitor the temperature rate and it increases the rate of diagnosis.[20,21] Capturing of location, image of face and monitoring of temperature can be done by Smart Helmet[22] and with Smart Glasses.[23] These devices also have the capability to make interaction of human very less. Electronic ankle bracelet is the example of Q-Band[24] which was used in USA for monitoring quarantine cases and it is very cost effective. For monitoring social distance by the people, EasyBand[25] play an important role during COVID. Instant Trace is one of the types of proximity Trace[26] device which is used for tracing the acquaintances of infected person in an organization device.

Drones also play a vital role during pandemic. These are just like aeroplane which consists of GPS, cameras, and sensors and where human activity is very less.[27] To capture the temperature in crowd and to reduce human interaction warm imaging drone is used.[28] To sanitize infected areas and to avert the health workers from infection sanitizer drone[29] plays an effective role in pandemic. Clinical robot[30] reduces the visit of person in hospitals and it helped to increase the convenience of treatment. For monitoring of crowd, infected areas, multipurpose drone, and observation drones are used.[31,33] To broadcast the COVID-related information declarement drone[32] is used in pandemic.

In Pandemic the real worth of various applications of smartphone was utilized by the people in year 2020. This year shows that almost 3.5 billion smartphones are active and the peoples are utilizing various apps to know more about pandemic.[34,35] Retail, farming, medical filed, etc. are various fields where smartphone applications work exceptionally in a proficient manner.[36,37] Various applications such as nCapp, Stop Corona, Aarogya Setu, etc. are utilized by the governments, organizations, and people all over the world. To control the health of patients in long term nCapp[38] is used and in USA for COVID test at low cost various people used a kit which was connected to smartphone application Detecta Chem[39] and to get daily reports of patients such as location, symptoms and to build a record of high-risk spots Stop Corona[40] apps is utilized. AargoyaSetu app connects the people and health sector in an efficient manner[45] and is widely used in India. Russia developed Social Monitoring application to diagnose the COVID-19 patients and assess their information,[41] Selfie application[42] was developed by Poland for monitoring the patients. Civitas,[43]

StayHomeSafe,[44] TraceTogether[46] are used by the government in Singapore to notify the people which were very close to the infected person. Similarly, Hamagen,[47] Coalition,[48] BeAware Bahrain,[49] and eRouska[50] are the various applications which were utilized for providing best services to the patients and governments all over the world. Whatsapp[51] is one of the wonderful applications which was widely used by many countries during pandemic for providing health care support.

The major requirement of IOT devices in these crucial periods will definitely unwrap the vision of trailblazers, scientists, and designers to IOT innovation, which brings noteworthy development in this era after COVID. IOT was recognized chiefly for a lot more extensive scale idea, like savvy urban areas, shrewd vehicles, and so forth. This article featured how various ventures, which were never matured pre-COVID-19, would be benefitted from IOT after COVID. The utilization of IOT enjoys its benefits and impediments. A huge improvement might be seen soon with perceptible development in IOT research, development, and application in 2020. A portion of the IOT use cases will be exclusively motivated because of the necessity to oversee and adjust to the COVID-19 pandemic, for example, get in touch with following and self-teaching. In the year 2020[38] the use of 20 billion gadgets and web-associated things was normal and after COVID period IOT has rapidly become one of the most recognizable articulations across business and innovation and the innovation worldwide market is relied upon to develop to around 1.6 trillion dollars in market income.[39] Coordination of IOT with various advancements, for example, distributed computing and implanting actuators and shrewd sensors work with cooperation with savvy things, permitting simple entry in various areas, upgrading information trade productivity, and further developing stockpiling and processing power.[40] There is an increment toward IOT because of COVID and the recent pandemic that influences the anticipated evaluations of IOT development. The effect of change welcomed in 2020 is driving associations to progressively involve IOT advances for functional versatility.[45] Soon, we figure we will see a greater amount of IOT and its ability as it plays a significant variable during this pandemic. With developing advancements, for example, 5G that Future Internet 2021, 13, 105 15 of 24 can expand the capacities of IOT much more with quicker organizations and information.

3.3   CHALLENGES TO DEPLOY IOT

Practical implementation of Internet of Things (IOT) presents various challenges and these are discussed in this section.

Implementation of IOT needs consumption of huge amount of sensors, actuators, and processors in the space of interest, in order to monitor the events of interest and take appropriate actions. Further, these must be able to communicate among them and with a centralized system through a wireless network. The integrated devices, consisting of computing devices (processors and memory) wireless communication devices (Radio), and I/O interfaces for connecting needed sensors and actuators, are called motes and the wireless network of these devices is called wireless sensor and actuator networks (WSAN). These networks of unprecedented quality and scale monitor the space all the 24 h. The deployed devices called nodes are battery powered, small in size, and unobtrusive. Often a network may need to have hundreds to thousands of nodes deployed. to perform its function. The main challenges in building these networks are as under:

- We do not have experience of deploying more than a hundred nodes at this time. The performance of the system when it is scaled up is to be studied. Often the battery lives are small (a few months) and maintenance of batteries in a large network, often in difficult terrain, is going to be a difficult task. Most deployments tend to be in the harsh environments (the devices are exposed to sun, wind, rain, vandalism, etc.) and the devices and networks are to be maintained in those environments.

- Present Motes are large in size and are expensive for ubiquitous deployment. The deployments are to be handcrafted today and we need to automate the process of deployment.

- Mote radio ranges are limited and we need more energy efficient radio for larger range. Further we need better protocols and algorithms for more efficient network operation.

- To diminish the obstruction between the human's cognitive model of what they crave to achieve and the computer's understanding of the user's task, a systems is required. So the systems we design need to mimic human–human interactions—spoken and written words, visual detection and identifying objects, visual communication, gestures, capacity, smell, and taste. Can we communicate through our thoughts (Brainwaves)? We have to go a long way to seamlessly integrate the networks into the human activities.

- The systems that we are visualizing must exhibit adequate intelligence and we have to incorporate this artificial (non-natural, manufactured) intelligence into them. The intelligent behavior is computation intensive and so we need more powerful and smaller computing systems to implement intelligent behavior.

- To build smart spaces, we need to embed computing infrastructure extensively in our physical infrastructure—objects around us, buildings and roads so that the smart space brings together the two worlds (computing and physical worlds) that have been disjoint till now. For this we need smaller and more powerful computing infrastructure.

- Extensive and context sensitive profiling of human beings is needed, and this information is to be provided to the system for intelligent decisions and treating the system as an extension of human being. In this venture we have to preserve the privacy of people and that needs a delicate balance between what profile of the human being is to be provided to the system and what should not be provided.

- Complete vision of IOT is achieved when computing technology is embedded into objects that we deal with and does not stand out. In such a case we interact with objects in a normal way and we do not see computers but only objects around us and that is the idea from a user's view and consciousness leading to minimal distraction of the user—no surprises for us.

- The intensity of interactions between a user's invisible personal computing space and his/her environment increases when the smart spaces grow in sophistication. This has severe communication bandwidth, energy consumption of the systems, and distraction implications for a user using wireless communication systems. The presence of multiple users will further complicate this problem. This is to be addressed by limiting interactions with unwanted (logically distant) entities.

- In the society the smartness of the objects varies over a wide range and depends upon the context. The large dynamic range of "smartness" can be a problem in making this technology invisible and also it is an implementation challenge. The level of "smartness" we can build into space around us differs. The smartness of conference room or office may be different from that of average home and that of a city road would differ from a village road. We need to address the problem. We also see more smartness close to us and it goes on decreasing with distance. Can we hide the smartness as the object is far and bring it into focus only when it is close to us?

- All the systems under IOT need to know the user intent in order to respond to him/her in a meaningful manner. To know the user's intent one may require user profile. For example when the user is to be hospitalized, the user may have his own choice of the hospital depending upon his/her preferences which might change with time, type of emergency, type of the health problem, cost level of the hospital, etc. Pervasive systems need to consider all these factors and take a decision.

- IOT systems need to know the user profile, intent, and behavior to provide proactive services. This means that it anticipates and knows all needed information about the user. Does this not lead to loss of privacy of the user as the system knows all information of the user? It knows our likes, dislikes, our preferences, out intentions, our behavior, etc. How do we protect this information from being exploited by others? How do we provide security to the user data?

- If IOT systems are to be used by all the population, society would have to provide extensive IOT computing resources as it provides electricity and water at all the places! When there are inadequate resources the human being and the pervasive system must adapt to the situation. IOT system may allocate resources as it perceives the priority of the problem, or the user may request the system for a certain level of resources. So dynamic resource management would be a major challenge. Memory management, efficient use of battery, controlling the level of accuracy, slower delivery, and amount of resource allocated for the service are some techniques that could be used for IOT.

- Capabilities such as proactivity and self-tuning would lead to increased energy demand for the pervasive systems.

- Context awareness (human and environmental contexts) needs capturing the contexts. The question is how do we capture rich contexts such as emotions, physiological conditions and how are they represented in the system? How does this system retain and update this info? How does it use the info?

3.3.1 SECURITY AND IOT

Tragically, most of these gadgets and applications are not intended to deal with the security and protection assaults and it builds a great deal of safety and protection issues in the IOT organizations, for example, privacy, confirmation, information trustworthiness, access control, mystery, and so forth.[53] In IOT, every one of the gadgets and individuals are associated with one another to offer types of assistance whenever needed and at any spot. For the IOT, some security necessities should be satisfied to keep the organization from vindictive assaults.[52–55]

Here, probably the most required capacities of a solid organization are momentarily examined.

- **Versatility to assaults:** The framework should be sufficiently able to recuperate itself on the off chance that assuming it crashes during information transmission. For a model, a server working in a multiuser climate, it should be astute and sufficiently able to shield itself from gatecrashers or a busybody. For the situation, on the off chance that it is down it would recuperate itself without suggestion of the clients of its down status.
- **Client protection:** The information and data should be in safe hands. Individual information ought to just be gotten by approved individuals to keep up with the client protection. It implies that no insignificant confirmed client from the framework or some other sort of client cannot approach the private data of the client.

3.3.2 ANALYSIS OF THREAT IN IOT

As per the situation of IOT, various security threats which include domain name system security threats, HTTPs threats are analyzed. In this section, various security threats are discussed which occur by using IOT system.

- **Directing assault:** Routing data in IOT can be satirized, changed, or replayed, to make steering circles, assaults, and so forth.
- **Forswearing of service assault:** Typically, things have tight memory and restricted calculation; they are along these lines defenseless against asset depletion assault. Aggressors can consistently send solicitations to be handled by explicit things to exhaust their assets.

This is particularly hazardous in the IOTs since an assailant may be situated in the backend and target resource-constrained gadgets in an LLN. Furthermore, DoS assault can be sent off by actually sticking the correspondence channel, subsequently separating the T2T correspondence channel. Network accessibility can likewise be upset by flooding the organization with an enormous number of bundles.

3.4 SECURITY SYSTEM

The term security subsumes a wide scope of various ideas. In any case, it alludes to the essential arrangement of safety administrations including privacy, confirmation, uprightness, approval, non-renouncement, and accessibility. These security administrations can be carried out through various cryptographic instruments, for example, block figures, hash capacities, or mark calculations. For every one of these instruments, a strong key administration foundation is essential to dealing with the necessary cryptographic keys. This segment gives a security way to deal with an IP-based network. We utilize the accompanying wording to investigate and arrange security perspectives in the IOT.[56–58]

- Bootstrapping incorporates the validation and approval of a gadget as well as the exchange of safety boundaries taking into consideration confided in activity.
- Network security depicts the components applied inside an organization to guarantee confided in activity of the IOT. Network security can incorporate various components going from secure steering to information connect layer and organization layer security.

3.4.1 CUTTING EDGE OF IOT SECURITY

- Establishment of reliance: In most IOT situations trust should be laid out impromptu with beforehand unregistered and obscure companions, and lacking of user communication. There is a requirement of novel methodology which can establish the foundation of trust. Present Trust programmes are not satisfying the requirement of user.
- Blockchain and IOT: The conventions depending upon block chain strategy are acquiring prevalence which can deal with the

test of laying out reliance. The most significant structural squares of the future are that blockchain-based smart contracts can be used for IOT-based foundations. Because they play a vital role for business-basic collaboration linking gadgets without direct human association. In any case, blockchains require computational assets and have high data transfer capacity upward.

- Trust in Platforms: Two methodologies on computerized foundation of confidence in distant stages exist: equipment and programming far off confirmation. Extra asset utilization by such equipment is not satisfactory for some battery-controlled gadgets. Programming far off verification can give an OK security level to most applications, yet it cannot thoughtfully ensure reliance in the general stage. The auxiliary advancement of system obscurity, cryptography of white-box, and control-stream uprightness innovations can give comprehensive programming just remote attestation. Additionally equipment is not OK for some battery-controlled gadgets. Programming far off confirmation can give a satisfactory assurance level to most applications yet it cannot theoretically ensure trust in the general stage. Further improvement of code confusion, white-box cryptography, and control-stream honesty advances can give comprehensive programming just distant validations from now on.
- Personality Management: Later on, independent information trades among various substances are relied upon to be controlled in light of cutting edge security and trust the executives advancements, for example use control.
- Security: Data use control is an expansion of conventional right to use control ideas. Future developments in information usage control will expand on traditional access control concepts to track and mark information as it is handled by various systems. It will characterize fine-granular utilization limitations to implement security properties over huge informational indexes while as yet taking into consideration running learning calculations and investigation over them.

3.5 CONCLUSION

While the world is battling with the COVID-19 pandemic, numerous innovations have been carried out to battle against this sickness. One of these advancements is the Internet of Things (IOT), which has been generally utilized in the medical care industry. During the COVID-19 pandemic, this innovation has shown exceptionally uplifting results managing this sickness. For this chapter, we led a study on the new proposed IOT gadgets expecting to help medical care laborers and specialists during the COVID-19 pandemic. We looked into the IOT-related innovations and their executions in three stages, including "Early Diagnosis," "Quarantine Time," and "After Recovery." For each stage, we assessed the job of IOT-empowered/connected in battling COVID-19. IOT innovation can be very effective for this pandemic; however, it is likewise basic to think about the security of information. By carrying out IOT innovation appropriately in a solid manner, more patients, with inner serenity, can partake in their treatment utilizing IOT gadgets. Accordingly, specialists and medical care laborers can all the more likely answer to pandemics. Therefore, the effect of these kinds of illnesses, including contaminations, hospitalizations, and passing rate, can be fundamentally diminished.

KEYWORDS

- pandemic
- COVID-19
- security
- IoT
- security systems

REFERENCES

1. Z. H.; Ali, H. A.; Badawy, M. M. Internet of Things (IOT): Definitions, Challenges and Recent Research Directions. *Int. J. Comput. Appl.* **2015,** 128 (1), 37–47.
2. HaddadPajouh, H.; Dehghantanha, A.; Parizi, R. M.; Aledhari, M.; Karimipour, H. A Survey on Internet of Things Security: Requirements, Challenges, and Solutions. *Internet of Things* **2019,** 3, 100–129.
3. Ashton, K.; et al. That 'Internet of Things' Thing. *RFID J.* **2009,** 22 (7), 97–114.
4. da Costa, C. A.; Pasluosta, C.F.; Eskofier, B.; da Silva, D. B, da Rosa Righi, R. Internet of Health Things: Toward Intelligent Vital Signs Monitoring in Hospital Wards. *Artif. Intell. Med.* **2018,** 89, 61–69.

5. Islam, S. M. R.; Kwak, D.; Kabir, M. D.H.; Hossain, M.; Kwak, K. S. The Internet of Things for Health Care: A Comprehensive Survey. *IEEE Access* **2015**, 3, 678–708.

6. Hu, F.; Xie, D.; Shen, S. In On the Application of the Internet of Things in the Field of Medical and Health Care, 2013 IEEE International Conference on Green Computing and Communications an IEEE Internet of Things and IEEE Cyber, Physical and Social Computing. *IEEE, 2013*; pp 2053–2058.

7. Qi, J.; Yang, P.; Min, G.; Amft, O.; Dong, F.; Xu, L. Advanced Internet of Things for Personalised Healthcare Systems: A Survey. *Pervasive Mob. Comput.* **2017**, 41, 132–149.

8. IOT in Healthcare Market. [Online] 2020. https://www.marketsandmarkets.com/Market-Reports/IOT-healthcaremarket-160082804.html.

9. Lovelace, Jr B. Scientists say the Coronavirus is at Least as Deadly as the 1918 Flu Pandemic 2020 [Online]. https://www.cnbc.com/berkeley-lovelace-jr/. (accessed Sept 5, 2020)

10. WHO (2020) Coronavirus Disease (COVID-19) [Online]. https://bit.ly/2ZU5x08 (accessed July 2020, 09)

11. Symptoms of Coronavirus 2020 [Online]. https://www.cdc.gov/coronavirus/2019-ncov/symptoms-testing/ symptoms.html. CDC (2020) Quarantine if you might be sick. https://www.cdc.gov/coronavirus/2019-ncov/ if-you-are-sick/quarantine.html.

12. Wang, W.; Tang, J.; Wei, F. Updated Understanding of the Outbreak of 2019 Novel Coronavirus (2019-nCoV) in Wuhan, China. *J. Med. Virol.* **2020**, 92 (4), 441–447.

13. Peeri, N. C.; Shrestha, N.; Rahman, M. S.; Zaki, R.; Tan, Z.; Bibi, S.; Baghbanzadeh, M.; Aghamohammadi, N.; Zhang, W.; Haque, U. The SARS, MERS and Novel Coronavirus (COVID-19) Epidemics, the Newest and Biggest Global Health Threats: What Lessons have we Learned? *Int. J. Epidemiol.* **2020.**

14. Singh, R. P.; Javaid, M.; Haleem, A.; Suman, R. Internet of things (IOT) Applications to Fight Against COVID-19 Pandemic, Diabetes & Metabolic Syndrome: Clinical Research & Reviews, 2020.

15. Ting, D. S. W.; Carin, L.; Dzau, V.; Wong, T. Y. Digital Technology and COVID-19. *Nat. Med.* **2020**, 26 (4), 459–461.

16. Hameed, S.; Khan, F. I.; Hameed, B. Understanding Security Requirements and Challenges in Internet of Things (IOT): A Review. *J. Comput. Netw. Commun.* **2019**, 2019, 1–14.

17. Rahmani, A. M.; Mirmahaleh, S. Y. H. Coronavirus Disease (COVID-19) Prevention and Treatment Methods and Effective Parameters: A Systematic Literature Review. *Sustain. Cities Soc.* **2021**, 64, 102568. [CrossRef]

18. Talavera, J. M.; Tobón, L. E.; Gómez, J. A.; Culman, M. A.; Aranda, J. M.; Parra, D. T.; Quiroz, L. A.; Hoyos, A.; Garreta, L. E. Review of IOT Applications in Agro-industrial and Environmental Fields. Comput. Electron. Agric. **2017**, 142, 283–297.

19. Chamola, V.; Hassija, V.; Gupta, V.; Guizani, M. A Comprehensive Review of the COVID-19 Pandemic and the Role of IOT, Drones, AI, Blockchain, and 5G in Managing its Impact. *IEEE Access* **2020**, 8, 90225–90265.

20. Suleman, H. How to Use the IOT to Keep Your Restaurant Clean and Safe. FoodSafetyTech [Online]. https:// foodsafetytech.com/column/how-to-use-the-IOT-to-keep-your-restaurant-clean-and-safe/ (accessed Apr 15, 2021).

21. Singh, S.; Hamidon, N.; Zuber, M.; Kamarul, A. Wireless Sensing Technology with IoMT Approach for Continuous Monitoring of Breathing Rate and Volume During COVID-19. *Front. Sustain. Cities* **2021,** 3, 6.

22. Uday, S.; Jyotsna, C.; Amudha, J. In Detection of Stress using Wearable Sensors in IOT Platform, Proceedings of the International Conference on Inventive Communication and Computational Technologies, ICICCT 2018, Coimbatore, India, 20–21 April 2018.

23. ABTraceTogether. Alberta.ca. [Online]. https://www.alberta.ca/ab-trace-together. aspx (accessed Mar 3, 2021).

24. Stanford Children's Health. About Telehealth Services (Virtual Visits)—Stanford Children's Health [Online]. https: //www.stanfordchildrens.org/en/telehealth/about-virtual-visits (accessed Feb 26, 2021).

25. Kinsa Health. Kinsa Smart Thermometers; Kinsa Inc. *Future Int.* **2021,** 13, 105. https://www.kinsahealth.co/products/ (accessed on Feb 26, 2021).

26. Isabella, M. A.; Seetha Lekshmi, K.; Thamizhvaani, E. P.; Vishali, S. IOT Based Emergency Medical Services. *Int. J. Eng. Tech.* **2018,** 4, 1–4.

27. Gupta, N.; Jilla, S. In Digital Fitness Connector: Smart Wearable System. Proceedings of the 1st International Conference on Informatics and Computational Intelligence, ICI 2011, Bandung, Indonesia, Dec 12–14, 2011.

28. Munawar, H.; Khan, S.; Qadir, Z.; Kouzani, A.; Mahmud, M. Insight into the Impact of COVID-19 on Australian Transportation Sector: An Economic and Community-Based Perspective. *Sustainability* **2021,** 13, 1276.

29. Gray, R. S. Agriculture, Transportation, and the COVID-19 Crisis. *Can. J. Agric. Econ.* **2020,** 68, 239–243.

30. Muthuramalingam, S.; Bharathi, A.; Kumar, S. R.; Gayathri, N.; Sathiyaraj, R.; Balamurugan, B. In IOT Based Intelligent Transportation System (IOT-its) for Global Perspective: A Case Study, Intelligent Systems Reference Library; Springer Science and Business Media LLC: Berlin/Heidelberg, Germany, 2019; vol 154.

31. Darsena, D.; Gelli, G.; Iudice, I.; Verde, F. Safe and Reliable Public Transportation Systems (SALUTARY) in the COVID-19 Pandemic. arXiv 2020.

32. Sutar, S.; Koul, R.; Suryavanshi, R. In Integration of Smart Phone and IOT for Development of Smart Public Transportation System, Proceedings of the 2016 International Conference on Internet of Things and Applications (IOTA), Pune, India, Jan 22–24, 2016; pp 73–78.

33. Al-Dweik, A.; Muresan, R.; Mayhew, M.; Lieberman, M. In IOT-Based Multifunctional Scalable Real-Time Enhanced Road Side Unit for Intelligent Transportation Systems, Proceedings of 30th Annual IEEE Canadian Conference on Electrical and Computer Engineering (IEEE 2017 CCECE), Windsor, ON, Canada, May 30–Apr 3, 2017

34. Gregory, J. The Internet of Things: Revolutionizing the Retail Industry [Online]. https://www.accenture.com/ _acnmedia/Accenture/ConversionAssets/DotCom/ Documents/Global/PDF/Dualpub_14/AccentureTheInternetOfThings. pdf (accessed Apr 21, 2021).

35. De Vass, T.; Shee, H.; Miah, S. J. IOT in Supply Chain Management: A Narrative on Retail Sector Sustainability. *Int. J. Logist. Res. Appl.* **2020,** 1–20.

36. Bashir, A.; Izhar, U.; Jones, C. IOT Based COVID-19 SOP Compliance Monitoring and Assisting System for Businesses and Public Offices [Online]. https://ecsa-7. sciforum.net/ (accessed Apr 22, 2021).

37. Petrovic, N.; Kocic, D. IOT-based System for COVID-19 Indoor Safety Monitoring. IcETRAN **2020,** 2020, 1–6.

38. Statista Research Department. IOT Market Size Worldwide 2017–2025; Statista, Jan 22, 2021. https: //www.statista.com/statistics/976313/global-IOT-market-size/ (accessed Feb 26, 2021).

39. Hung, M. Leading the IOT—Gartner Insights on How to Lead in a Connected World. *Gart. Res.* **2017,** 1, 1–5.

40. Domb, M. Smart Home Systems Based on Internet of Things. In IOT and Smart Home Automation [Working Title]; IntechOpen: London, UK, 2019; pp 25–37.

41. HTF Market Intelligence Consulting. IOT In Logistics Market May See a Big Move; Cisco Systems, IBM. https://www.openpr.com/news/2134501/IOT-in-logistics-market-may-see-a-big-move-cisco-systems-ibm (accessed Feb 26, 2021).

42. IOT Business News. IOT News—Cisco Predicts Rapid Growth in The IOT Logistics Market—IOT Business News [Online]. https://IOTbusinessnews. com/2020/08/19/06126-cisco-predicts-rapid-growth-in-the-IOT-logistics-market/ (accessed Feb 26, 2021).

43. Morrish, J. Global IOT Market to Grow to $1.5trn Annual Revenue by 2030— IOT Now News, How to Run an IOT Enabled Business [Online]. https://www. IOT-now.com/2020/05/20/102937-global-IOT-market-to-grow-to-1-5trn-annual-revenue-by-20 30/ (accessed Feb 26, 2021).

44. Horwitz, L. Internet of Things (IOT)—The Future of IOT Miniguide: The Burgeoning IOT Market Continues, Cisco [Online]. https://www.cisco.com/c/en/us/solutions/internet-of-things/future-of-IOT.html (accessed Feb 26, 2021).

45. Medberry, P. Industrial IOT: Top 3 Trends for 2021, Cisco Blogs [Online]. https:// blogs.cisco.com/internet-of-things/ industrial-IOT-top-3-trends-for-2021 (accessed Feb 26, 2021).

46. Uma, M.; Padmavathi, G. A Survey on Various Cyber Attacks and their Classification. Int. J. Netw. Secur. **2013,** 15, 5.

47. Abomhara, M.; Køien, G. M. Cyber Security and the Internet of Things: Vulnerabilities, Threats, Intruders and Attacks. *J. Cyber Secur. Mobil.* **2015,** 4, 65–88. [CrossRef]

48. Hoque, N.; Bhuyan, M. H.; Baishya, R.; Bhattacharyya, D.; Kalita, J. Network Attacks: Taxonomy, Tools and Systems. *J. Netw. Comput. Appl.* **2014,** 40, 307–324. [CrossRef]

49. Canongia, C.; Mandarino, R. Cybersecurity: The New Challenge of the Information Society. In Crisis Management: Concepts, Methodologies, Tools, and Applications; IGI Global: Hershey, PA, USA, 2013; vol 1–3.

50. Injac, O.; Šendelj, R. National Security Policy and Strategy and Cyber Security Risks. In Identity Theft: Breakthroughs in Research and Practice; IGI Global: Hershey, PA, USA, 2016.

51. What Is Cyberspace? Definition from WhatIs.com. [Online]. https://whatis.techtarget. com/definition/cyberspace (accessed Mar 4, 2021).

52. Miorandi, D.; Sicari, S.; Pellegrini, F. D.; Chlamtac, I. Internet of Things: Vision, Applications and Research Challenges. *Ad Hoc Networks* **2012,** 10, 1497.

53. Schukat, M.; Castilla, P. C.; Melvin, H. In Trust and Trust Models for the IOT, Security and Privacy in Internet of Things (IOTs): Models, Algorithms, and Implementations; Hu, F., Ed.; CRC Press, 2016.

54. IOT 2020: Smart and Secure IOT Platform. IEC White Paper [Online]. http://www.iec.ch/whitepaper/IOTplatform

55. Lu, X.; Qu, Z.; Li, Q.; Hui, P. Privacy Information Security Classification for Internet of Things Based on Internet Data. *Int. J. Distrib. Sens. Netw.* **2015,** *11* (8), 932941.

56. Kanniappan, J.; Rajendiran, B. Privacy in the Internet of Things. In The Internet of Things in the Modern Business Environment; Lee, Ed.; IGI Global, 2017.

57. Practical BLE Throughput. Rigado LLC, 2016 [Online]. www.rigado.com/modules.

58. http://atmosphere.anaren.com/wiki/Data_rates_using_BLE.

59. Anitta, V.; Fincy, F.; Ayyappadas, P. S. Security Aspects in 6lowpan Networks. *IOSR J. Electron. Commun. Eng.* **2015,** 10, 8.

CHAPTER 4

MOBILE APP DEVELOPMENT PRIVACY AND SECURITY CHECKLIST DURING COVID-19

HENA IQBAL, TANJINSIKDER, YASMIN ALHAYEK, NOORA ALFURAIS, and NUJUOM ASSAR

Department of Information Technology, Ajman University, Ajman, UAE

ABSTRACT

Apps for mobile devices have become an inextricable element of daily life. Because of the advancement of mobile technology, mobile access to high-speed internet, and the interactivity of mobile phone interfaces, they have dominated users' digital habits. Modern mobile phones include a wide range of potent essential sensors and attributes, such as a low-power Bluetooth sensor, special embedded sensors such as a digital navigation system, motion sensor, GPS module, wireless internet capabilities, recording device, humidity sensors, health-tracking sensors, and a camera, among other things. These value-added sensors have transformed people's lives in many ways. Such applications and features are proven effective from time to time for humans. Apart from being merit, this feature was put in the question of security they provide to the user. With time several applications were questioning their tracking policy. This study is based on the role of the mobile phone application in individual life and the privacy predictor of those applications. This study is conducted with the help of 356 responses being collected via in-person interactions with the mobile

The Fusion of Artificial Intelligence and Soft Computing Techniques for Cybersecurity.
M. A. Jabbar, Sanju Tiwari, Subhendu Kumar Pani, & Stephen Huang (Eds.)

users in UAE (all seven Emirates). From the study, it has emerged that there is significant association between the demographic factors, age, gender, occupation, income, and educational qualification, with respect to three different variables, concerning the security and privacy of mobile apps during COVID-19.

4.1 INTRODUCTION

Mobile phones have become a central part of our day-to-day life.[8] In recent years, mobile phones have seen an upsurge in their usage. With time mobile phone companies are updating and modifying themselves to meet the changing demands of their user for improved performance in everyday lifestyle. The first-ever mobile phone was introduced back in 1973 by Martin Cooper, an engineer for Motorola. In the earlier stage of the mobile phones, it was used to place calls, texts, and other minor stuff. In 2007, Steve Jobs brought a revolution in the mobile industry by showcasing the first-ever smart mobile device popularized as iPhone. At that time, this smartphone allows its user to use numerous applications with supportable multitasking. The entire mobile phone industry felt the need to modify themselves to survive in the market by fulfilling customer's requirements and they started enrolling more smartphone devices. Currently, these devices have evolved so much that it has become a part of daily life. Apart from basic tasks, these devices have become a medium of entertainment and distance communication. Application developers have molded such applications frequently to make their users stick to them in a long run.[7,41]

As per Statista, 2018, on an average 6100 applications were being uploaded by the google play store for the users to make use of, between 2016 and 2018. However, the usage rate of these applications is negligible and most of the applications are left unused or uninstalled after a short span of use, while many applications are used for not more than 3 to 4 months on an average.[21,36] With rising concerns regarding these applications, it was found in further research that the threat to the privacy and security of confidential data of the users were the pivotal reason behind such usage behavior among mobile users.[19,32,42] Usage of newly uploaded applications for the users depends on their concerns related to that application.[19] Users hesitate to make use of any unsecured application as they fear that these apps may hold malicious cryptography which can cause harm to their confidential data, can track their activities, and can hack their information.[30] It highlights the need for improved security along with timely modifications to keep a check on the possible vulnerabilities. However, the security of the users cannot be maintained only by improved security but also by time-to-time scrutinization of malicious activities of different users, restricted release and usage of unsecured applications, and so on.[37,38]

The rise in usage behavior of any application by a user is directly dependent on their perception of that application. If a user possesses a positive perception of that application, then they are more likely to download and use that application for a long and vice versa.[19] In 2015, Starbucks admitted that a hacker has hacked the details of their customers via their official application which resulted in fear among users or the customers reading their data and they end up uninstalling the application of Starbucks to safeguard their data. Similarly, in 2019 and again in 2021, Facebook (now Meta) faced the same when the personal information of millions of users was exposed; as a result, it lost its number of users at the threat to their data.[18,33,46] Hence, mobile application developers must understand and build a safe and secured application for their users, thereby gaining their trust in it.[9] Studying the perception and usage behavior of the users will help the developers to understand the concerns of their users well and provide them with the best and most secure applications and solutions.

4.2 REVIEW OF LITERATURE

Information and Communication Technologies (ICT) are a blessing to humans. They eased the exchange of words, information, and data than before. One of its biggest applications was seen during the pandemic COVID-19, which made people worldwide rely on ICT, AI, machine learning, and other technological gifts. The entire world was shifted to virtual platforms and started working remotely. It made government as well as various technological developers come up with more innovative and secure ways to tackle such a situation.[45] Innovative methods and tools were developed to make life smoother during the pandemic, that is many mobile-based applications were developed to keep people aware of the going scenario around the world, to track the spread of the virus around the world, and so on, to help them to take timely measures and required precautions to keep themselves safe and secured.

With time, this became the topic of research for scholars and researchers around the world. Besides this, one of the critical aspects which was the crux of all the debate was the security and privacy of the mobile applications being released and used during the pandemic COVID-19. Various scholars highlighted the need for safe and secured measures to keep the users at bay from any vulnerability and malicious applications.[10,44,50] Literature comprises studies based on the operations of mobile applications and their associated threats and vulnerabilities to them.[4,16,47] During COVID-19, many launched applications were based on contact-tracing features which included GPS, Bluetooth, and other wireless technologies, as used in maps and social networking sites.[3,14,39,40,43,52] This in turn made regulators like technological developers, governmental bodies, Federal Trade Commission, European Union Commission, and others inspect and explore various ways and methodologies to improve and develop the operations and security of various mobile applications.[5,20] Additionally, preference, perception, and usage behavior of the mobile users were also studied to understand the threats they faced to provide them with the best security and privacy protocols, thereby creating a safe and secure technology platform and environment for the users to use confidently.[26,48] Users' behavior was researched in various aspects concerning mobile applications[19,24] and the security provided by them, which ranges from computer security,[25] online security,[11] online purchase, and transactions.[12,28,34]

Mobile-based apps require various permissions and access to certain data to operate properly, like permission and access to use storage, gallery, microphone, camera location, and so on.[17] This, in turn, exposes the personal and confidential data to such mobile-based applications which can be misused or hacked by any malicious or vulnerable activities.[5] When a user grants permission of personal data to any application it becomes available to different insurance and advertisement companies, the public, and other unauthorized entities, which pose a threat to the personal and confidential data of the users.[1,2,29,53] In most cases, apps ask for permissions that are not required or not related to their basic operations, which are mostly unsecured apps or malicious dealers.[27,35] Many researchers study various aspects of mobile-based applications, the method of their operation, and the permissions that they ask for from their users.[6,13,16,22,23,31,51] These studies act as an aid for the applications developers to analyze and understand the needs of their users, possible threats and insecurities faced by them, and possible course of action or solutions that they can make use of to provide their users with safe and secured applications to make use of.

4.3 OBJECTIVE

To understand the association between the demographic factors of the mobile users to the privacy and security of mobile apps during the COVID-19.

4.4 RESEARCH METHODOLOGY

Structured questionnaire was incorporated to collect the responses for the study and casual face-to-face interactions were held with the mobile users in various parts of UAE. Samples were collected from Ajman, Dubai, Sharjah, Abu Dhabi, Fujairah, Ras Al Khaimah, and Umm Al Quwain. The data were collected from the mobile users of different age group, gender, occupation, income, and educational qualification. A total of 356 responses were collected for the study via various parameters. Each parameter studied for the survey was calculated using a five-point like a scale (from 1-strongly disagree, to 5-strongly agree).

A Chi-Square test is conducted to make a comparison between the expected output and the actual output. It helps to understand that whether the difference between the expected output and the actual output is by a fluke or is there any relationship between the variables to be studied. Hence, we used the chi-square test to find out the significant association between the demographic factors of the mobile users with respect to the security and privacy of mobile apps during COVID-19.

Motivating Variables:

- V1—Downloading an app from the web causes more security issues than websites.
- V2—Unsecured mobile apps provide bad authorization and authentication of the sensitive data which allows hackers to steal it.
- V3—One should avoid saving confidential credentials while using a mobile application.

4.5   FINDINGS AND ANALYSIS

4.5.1   GENDER

H_0—There is no significant association between the gender of the mobile users and the security and privacy of mobile apps during COVID-19.

H_1—There is no significant association between the gender of the mobile users and the security and privacy of mobile apps during COVID-19.

TABLE 4.1   Chi-Square Tests.

	Value	Df	Asymp. sig. (2-sided)
Pearson chi-square (V1)	71.700	3	.002
Likelihood ratio (V1)	71.500	3	.003
Linear-by-linear association (V1)	38.270	1	.002
N of valid cases (V1)	356		
Pearson chi-square (V2)	56.700	4	.002
Likelihood ratio (V2)	65.390	4	.002
Linear-by-linear association (V2)	0.860	1	.440
N of valid cases (V2)	356		
Pearson chi-square (V3)	65.015	4	.003
Likelihood ratio (V3)	73.350	4	.002
Linear-by-linear association (V3)	0.025	1	.890
N of valid cases (V3)	356		

Table 4.1, of Chi-Square, represents that the chi-square value of the factor V1 is 71.700 and the significant value is .002 which is less than 0.05, whereas the chi-sq. value of the factor V2 is 56.700 and the significant value is 0.002 which is less than 0.05; similarly, the chi-square value of the factor V3 is 65.015 and the significant value is 0.003 which is less than 0.05. This implies that the null hypothesis (H_0) is rejected and the alternative hypothesis (H_1) is accepted, that is gender of the mobile users has a significant association with the security and privacy of mobile apps during COVID-19, for variables V1, V2, and V3.

Females were found to be agreeing more with the possible threats to their confidential data while using a mobile app. They believe that downloading mobile apps from the web causes greater security threats than websites. This is as a result of mobile applications gathering and accessing their users' personal data to store and use for the future references, which resulted in women being more cautious about these mobile apps. Similarly, illegal authorization and hacking are also a huge threat to users' personal information. Women are also seen to have lesser trust in saving confidential data on a mobile phone, to avoid unlawful activities with their data, or their personal photos, contacts, emails, etc.

TABLE 4.2   Symmetric Measures.

		Value	Approx. sig.
Nominal by nominal	Contingency coefficient (V1)	0.415	.002
N of valid cases		356	
Nominal by nominal	Contingency coefficient (V2)	0.400	.002
N of valid cases		356	
Nominal by nominal	Contingency coefficient (V3)	0.407	.003
N of valid cases		356	

Table 4.2 of Symmetric Measure, represents that the Contingency Coefficient value of the factor V1 is 0.415, for the factor V2 it is 0.400, while for the factor V3 it is 0.407 which implies that the association between gender of the mobile users and the security and privacy of mobile apps during COVID-19 is medium with the variables V1, V2, and V3.

4.5.2   AGE

H_0—There is no significant association between the age of the mobile users and the security and privacy of mobile apps during COVID-19.

H_1—There is a significant association between the age of the mobile users and the security and privacy of mobile apps during COVID-19.

TABLE 4.3   Chi-Square Tests.

	Value	Df	Asymp. sig. (2-sided)
Pearson chi-square (V1)	65.223	9	.002
Likelihood ratio (V1)	55.609	9	.002
Linear-by-linear association (V1)	0.108	1	.750
N of valid cases (V1)	356		
Pearson chi-square (V2)	26.558	12	.007
Likelihood ratio (V2)	37.244	12	.003
Linear-by-linear association (V2)	0.002	1	.980
N of valid cases (V2)	356		
Pearson chi-square (V3)	39.670	12	.001
Likelihood ratio (V3)	46.862	12	.002
Linear-by-linear association (V3)	0.055	1	.820
N of valid cases (V3)	356		

Table 4.3, of Chi-Square, represents that the chi-square value of the factor V1 is 65.223 and the significant value is 0.002 which is less than 0.05, whereas the chi-sq. value of the factor V2 is 26.558 and the significant value is 0.007 which is less than 0.05; similarly, the chi-square value of the factor V3 is 39.670 and the significant value is 0.001 which is less than 0.05. This implies that the null hypothesis (H_0) is rejected and the alternative hypothesis (H_1) is accepted, that is the age of mobile users has a significant association with the security and privacy of mobile apps during COVID-19, for variables V1, V2, and V3.

Threats related to privacy and safety of the data have caught the attention of people in recent times since the current generation is more aware of the possible threats to their data; therefore, they were found to be agreeing more with the possible threats to their confidential data while using a mobile app. Users aged 18–25 were found to trust websites more than applications because they are quite aware of the security threats arising out of them as these apps seek permissions and access to personal data of the users to start using it. They are also found to avoid unsecured applications because they can access private information, and can hack their important data. The younger generation also tend to not to save their private information on their mobile phones, to avoid the misuse of their confidential data.

TABLE 4.4   Symmetric Measures.

		Value	Approx. sig.
Nominal by nominal	Contingency coefficient (V1)	0.400	.002
N of valid cases		356	
Nominal by nominal	Contingency coefficient (V2)	0.300	.007
N of valid cases		356	
Nominal by nominal	Contingency coefficient (V3)	0.350	.001
N of valid cases		356	

Table 4.4 of Symmetric Measure, represents that the contingency coefficient value of the factor V1 is 0.400, for the factor V2 it is 0.300, while for the factor V3 it is 0.350 which implies that the association between the age of the mobile users and the security and privacy of mobile apps during COVID-19 is medium with the variable V1, while low with the variables V2 and V3.

4.5.3   OCCUPATION

H_0—There is no significant association between occupation of the mobile users and the security and privacy of mobile apps during COVID-19.

H_1—There is a significant association between occupation of the mobile users and the security and privacy of mobile apps during COVID-19.

TABLE 4.5   Chi-Square Tests.

	Value	Df	Asymp. sig. (2-sided)
Pearson chi-square (V1)	35.208	6	.002
Likelihood ratio (V1)	32.915	6	.003
Linear-by-linear association (V1)	8.415	1	.003
N of valid cases (V1)	356		
Pearson chi-square (V2)	110.886	8	.002
Likelihood ratio (V2)	97.735	8	.002
Linear-by-linear association (V2)	3.839	1	.058
N of valid cases (V2)	356		
Pearson chi-square (V3)	141.175	8	.003

TABLE 4.5 *(Continued)*

	Value	Df	Asymp. sig. (2-sided)
Likelihood ratio (V3)	158.605	8	.004
Linear-by-linear association (V3)	0.859	1	.360
N of valid cases (V3)	356		

Table 4.5 of Chi-Square represents that the chi-square value of the factor V1 is 35.208 and the significant value is 0.002 which is less than 0.05, whereas the chi-sq. value of the factor V2 is 110.886 and the significant value is 0.002 which is less than 0.05; similarly, the chi-square value of the factor V3 is 141.175 and the significant value is 0.003 which is less than 0.05. This implies that the null hypothesis (H_0) is rejected and the alternative hypothesis (H_1) is accepted, that is occupation has a significant association with the security and privacy of mobile apps during COVID-19, for variables V1, V2, and V3.

Users from various occupations, except service and business sector, were found to be more careful about security interference to their confidential data via apps on their mobile phones. The "Other" category of occupation includes students, small business owners, etc., and since it mostly consists of the younger generation, they are reportedly found to be more aware of the possible threats to their confidential data, being imposed by unsecure mobile applications. Hence, they believe websites are safer than unsecured mobile applications. Students keep various personal data on their mobile phones, which can be illegally accessed by hackers through unsecure applications. And thus, they avoid storing their private data on their cell phones in most of the cases.

TABLE 4.6 Symmetric Measures.

		Value	Approx. sig.
Nominal by nominal	Contingency coefficient (V1)	0.315	.002
N of valid cases		356	
Nominal by nominal	Contingency coefficient (V2)	0.499	.002
N of valid cases		356	
Nominal by nominal	Contingency coefficient (V3)	0.546	.003
N of valid cases		356	

Table 4.6 of Symmetric Measure represents that the contingency coefficient value of the factor V1 is 0.315, for the factor V2 it is 0.499, while for the factor V3 it is 0.546 which implies that the association between gender of the mobile users and the security and privacy of mobile apps during COVID-19 is low with the variable V1 while it is medium with the variables V2 and V3.

4.5.4 EDUCATIONAL QUALIFICATION

H_0—There is no significant association between educational qualification of the mobile users and the security and privacy of mobile apps during COVID-19.

H_1—There is a significant association between educational qualification of the mobile users and the security and privacy of mobile apps during COVID-19.

TABLE 4.7　Chi-Square Tests.

	Value	df	Asymp. sig. (2-sided)
Pearson chi-square (V1)	131.215	9	.003
Likelihood ratio (V1)	108.689	9	.005
Linear-by-linear association (V1)	2.755	1	.095
N of valid cases (V1)	356		
Pearson chi-square (V2)	191.720	12	.002
Likelihood ratio (V2)	123.565	12	.004
Linear-by-linear association (V2)	4.155	1	.049
N of valid cases (V2)	356		
Pearson chi-square (V3)	153.957	12	.004
Likelihood ratio (V3)	159.815	12	.006
Linear-by-linear association (V3)	9.087	1	.007
N of valid cases (V3)	356		

Table 4.7 of Chi-Square represents that the chi-square value of the factor V1 is 131.215 and the significant value is 0.003 which is less than 0.05, whereas the chi-sq. value of the factor V2 is 191.720 and the significant value is 0.002 which is less than 0.05; similarly, the chi-square value of the factor V3 is 153.957 and the significant value is 0.004 which

is less than 0.05. This implies that the null hypothesis (H_0) is rejected and the alternative hypothesis (H_1) is accepted, that is educational qualification of the mobile users has a significant association with the security and privacy of mobile apps during COVID-19, for the variables V1, V2, and V3.

Majority of the population in our country is graduated; therefore, most of our citizens are aware enough to care about their personal data. Graduates are more cautious about using unauthorized and unsecure applications, to avoid unlawful hacking of their data. They also elude loading all their private data on their cell phones for long or giving any access to these apps.

TABLE 4.8   Symmetric Measures.

		Value	Approx. sig.
Nominal by nominal	Contingency coefficient (V1)	0.525	.003
N of valid cases		356	
Nominal by nominal	Contingency coefficient (V2)	0.598	.002
N of valid cases		356	
Nominal by nominal	Contingency coefficient (V3)	0.557	.004
N of valid cases		356	

Table 4.8 of symmetric measure represents that the contingency coefficient value of the factor V1 is 0.525, for the factor V2 it is 0.598, while for the factor V3 it is 0.557 which implies that the association between educational qualification of the mobile users and the security and privacy of mobile apps during COVID-19 is medium with the variables V1, V2, and V3.

4.5.5   INCOME

H_0—There is no significant association between income of the mobile users and the security and privacy of mobile apps during COVID-19.

H_1—There is a significant association between income of the mobile users and the security and privacy of mobile apps during COVID-19.

TABLE 4.9   Chi-Square Tests.

	Value	df	Asymp. sig. (2-sided)
Pearson chi-square (V1)	77.324	12	.002
Likelihood ratio (V1)	75.227	12	.001
Linear-by-linear association (V1)	0.927	1	.349
N of valid cases (V1)	356		
Pearson chi-square (V2)	166.128	16	.001
Likelihood ratio (V2)	146.827	16	.003
Linear-by-linear association (V2)	31.329	1	.004
N of valid cases (V2)	356		
Pearson chi-square (V3)	137.974	16	.002
Likelihood ratio (V3)	150.817	16	.002
Linear-by-linear association (V3)	5.125	1	.027
N of valid cases (V3)	356		

Table 4.9, of Chi-Square, represents that the chi-square value of the factor V1 is 77.324 and the significant value is 0.002 which is less than 0.05, whereas the chi-sq. value of the factor V2 is 166.128 and the significant value is 0.001 which is less than 0.05; similarly, the chi-square value of the factor V3 is 137.974 and the significant value is 0.002 which is less than 0.05. This implies that the null hypothesis (H_0) is rejected and the alternative hypothesis (H_1) is accepted, that is income has a significant association with Mobile App Privacy and Security during COVID-19, for variables V1, V2, and V3.

Users from lower income group were found to be agreeing to the possible threats to their private data while using any unsecured apps in their mobile. Lower income users lack the privilege or they avoid to use premium services of any application, thereby restricting them to free application and their free services which can most of the time possess threat to the confidential data stored in the mobile of the users. As a result, lower income users were found to be more careful and protective toward their personal data and avoid using unsecured applications; they also keep limited personal data in their mobiles and avoid saving passwords and other such details in such apps in their mobiles.

TABLE 4.10   Symmetric Measures.

		Value	Approx. sig.
Nominal by nominal	Contingency coefficient (V1)	0.435	.002
N of valid cases		356	
Nominal by nominal	Contingency coefficient (V2)	0.570	.001
N of valid cases		356	
Nominal by nominal	Contingency coefficient (V3)	0.536	.002
N of valid cases		356	

Table 4.10 of Symmetric Measure represents that the Contingency Coefficient value of the factor V1 is 0.435, for the factor V2 it is 0.570, while for the factor V3 it is 0.536 which implies that the association between the income of the mobile users and the security and privacy of mobile apps during COVID-19 is medium with the variables, V1, V2, and V3.

4.5.6   MARITAL STATUS

H_0—There is no significant association between marital status of the mobile users and the security and privacy of mobile apps during COVID-19.

H_1—There is a significant association between marital status of the mobile users and the security and privacy of mobile apps during COVID-19.

TABLE 4.11   Chi-Square Tests.

	Value	df	Asymp. sig. (2-sided)
Pearson chi-square (V1)	14.447	3	.003
Likelihood ratio (V1)	19.115	3	.002
Linear-by-linear association (V1)	0.737	1	.385
N of valid cases (V1)	356		
Pearson chi-square (V2)	26.545	4	.002
Likelihood ratio (V2)	37.254	4	.004
Linear-by-linear association (V2)	0.003	1	.989
N of valid cases (V2)	356		

Table 4.11, of Chi-Square, represents that the chi-square value of the factor V1 is 14.447 and the significant value is 0.003 which is less than 0.05, whereas the chi-sq. value of the factor V2 is 26.545 and the significant value is 0.002 which is less than 0.05. This implies that the null hypothesis (H_0) is rejected and the alternative hypothesis (H_1) is accepted; that is the marital status of the mobile users has a significant association with the security and privacy of mobile apps during COVID-19, for variables V1 and V2.

Unmarried individuals are seen to be more aware of their data privacy. These individuals were found to keep range personal data in their mobiles, that they fear would get hacked easily causing huge loss to them. Therefore, they avoid using applications, and choose websites instead and they avoid keeping and saving any confidential data in their mobile phones or in the apps they use to keep their data safe and secure.

TABLE 4.12　Symmetric Measures.

		Value	Approx. sig.
Nominal by nominal	Contingency coefficient (V1)	0.201	.003
N of valid cases		356	
Nominal by nominal	Contingency coefficient (V2)	0.278	.002
N of valid cases		356	

Table 4.12 of Symmetric Measure represents that the contingency coefficient value of the factor V1 is 0.201, while for the factor V2 it is 0.278, which implies that the association between the marital status of mobile users and the security and privacy of mobile apps during COVID-19 is low with the variables V1 and V2.

4.6　CONCLUSION

In our networked world, the rise of mobile devices has resulted in a plethora of applications that serve a variety of objectives. Today's modern lifestyle is increasingly dependent on mobile apps that provide a variety of services, including military applications, critical business services, banking, entertainment, and other diverse tasks. During the COVID-19 epidemic, this technology played a vital role in assisting individuals.

Several programs that allow the user to locate any infected individual nearby have been launched. Although digital technologies can play a big role in addressing current pandemic concerns and limiting viral propagation, the usefulness and accuracy of these systems, however, are dependent on application design and user engagement. Therefore, mobile phone users are becoming more aware of protecting their data, and are taking the necessary steps to avoid misuse and spread of their data. It has been found that the female users are more concerned toward their confidential data as compared to the men; however, the younger generation like students are more alert about safeguarding their information, and follow secure steps to avoid it. Unmarried individuals are also more watchful about their data getting spread. This brings to light the inattentiveness individuals with occupations like business and service possess with their information security. More awareness among the developer as well the users should be raised to make use of secured platforms and to avoid any loss of their important financial or personal data.

KEYWORDS

- mobile
- sensors
- applications
- security
- privacy
- COVID-19

REFERENCES

1. Many Popular Android Apps Leak Sensitive Data, Leaving Millions of Consumers at Risk, 2017a [Online]. https://tinyurl.com/yb7hfjxr
2. Researchers Spot Thousands of Android Apps Leaking user Data through Misconfigured Firebase Databases, 2017b [Online]. https://tinyurl.com/ybjdrcth
3. Altuwaiyan, T.; Hadian, M.; Liang, X. In *EPIC: Efficient Privacy-Preserving Contact Tracing for Infection Detection*, 2018 IEEE International Conference on Communications (ICC); IEEE, May 2018; pp 1–6.

4. Au, K. W. Y.; Zhou, Y. F.; Huang, Z.; Lie, D. In *Pscout: Analyzing the Android Permission Specification*, Proceedings of the 2012 ACM Conference on Computer and Communications Security, Oct 2012; pp 217–228.

5. Azad, M. A.; Arshad, J.; Akmal, S. M. A.; Riaz, F.; Abdullah, S.; Imran, M.; Ahmad, F. A First Look at Privacy Analysis of COVID-19 Contact-Tracing Mobile Applications. *IEEE Int. Things J.* **2020**, *8* (21), 15796–15806.

6. Barrera, D.; Kayacik, H. G.; Van Oorschot, P. C.; Somayaji, A. In *A Methodology for Empirical Analysis of Permission-Based Security Models and its Application to Android*, Proceedings of the 17th ACM Conference on Computer and Communications security, Oct 2010; pp. 73–84. DOI: https://doi.org/10.1145/1866307.1866317

7. Balapour, A.; Sabherwal, R. Usability of Apps and Websites: A Meta-regression Study, 2017.

8. Balapour, A.; Reychav, I.; Sabherwal, R.; Azuri, J. Mobile Technology Identity and Self-efficacy: Implications for the Adoption of Clinically Supported Mobile Health Apps. *Int. J. Inf. Manag.* **2019**, *49*, 58–68. DOI: https://doi.org/10.1016/j.ijinfomgt.2019.03.005

9. Balapour, A.; Nikkhah, H. R.; Sabherwal, R. Mobile Application Security: Role of Perceived Privacy as the Predictor of Security Perceptions. *Int. J. Inf. Manag.* **2020**, *52*, 102063. DOI: https://doi.org/10.1016/j.ijinfomgt.2019.102063

10. Chan, J.; et al. Pact: Privacy Sensitive Protocols and Mechanisms for Mobile Contact Tracing, 2020. arXiv preprint arXiv:2004.03544.

11. Chen, Y.; Zahedi, F. M. Individuals' Internet Security Perceptions and Behaviors: Polycontextual Contrasts between the United States and China. *MIS Q.* **2016**, *40* (1), 205–222. https://doi.org/10.25300/misq/2016/40.1.09

12. Chellappa, R. K. Consumers' Trust in Electronic Commerce Transactions: The Role of Perceived Privacy and Perceived Security; Unpublished Paper; July 5, 2018; Emory University: Atlanta, GA, 2008. http://www.bus.emory.edu/ram/papers/secpriv.pd

13. Chia, P. H.; Yamamoto, Y.; Asokan, N. In *Is this App Safe? A Large Scale Study on Application Permissions and Risk Signals*, Proceedings of the 21st International Conference on World Wide Web, Apr 2012; pp 311–320. DOI: https://doi.org/10.1145/2187836.2187879

14. Danquah, L. O.; Hasham, N.; MacFarlane, M.; Conteh, F. E.; Momoh, F.; Tedesco, A. A.; Jambai, A.; Ross, D. A.; Weiss, H. A. Use of a Mobile Application for Ebola Contact Tracing and Monitoring in Northern Sierra Leone: A Proof-of-Concept Study. *BMC Infect. Dis.* **2019**, *19* (1), 1–12.

15. Enck, W.; Gilbert, P.; Han, S.; Tendulkar, V.; Chun, B. G.; Cox, L. P.; Jung, J.; McDaniel, P.; Sheth, A. N. Taintdroid: An Information-Flow Tracking System for Realtime Privacy Monitoring on Smartphones. *ACM Trans. Comput. Syst.* **2014**, *32* (2), 1–29.

16. Enck, W.; Ongtang, M.; McDaniel, P. In *On Lightweight Mobile Phone Application Certification*, Proceedings of the 16th ACM Conference on Computer and Communications security, Nov 2009; pp 235–245. DOI: https://doi.org/10.1145/1653662.1653691

17. Felt, A. P.; Greenwood, K.; Wagner, D. In *The Effectiveness of Application Permissions*, 2nd USENIX Conference on Web Application Development (WebApps 11), 2011.

18. Gross, A. Starbucks Data Breach Shows the Real Damage of a Breach, May 14, 2015. https://www.hipaasecurenow.com/index.php/starbucks-data-breachshows-real-damage-breach/.
19. Harris, M. A.; Brookshire, R.; Chin, A. G. Identifying Factors Influencing Consumers' Intent to Install Mobile Applications. *Int. J. Inf. Manag.* **2016**, *36* (3), 441–450. DOI: https://doi.org/10.1016/j.ijinfomgt.2016.02.004
20. Hekmati, A.; Ramachandran, G.; Krishnamachari, B. In *CONTAIN: Privacy-Oriented Contact Tracing Protocols for Epidemics*, 2021 IFIP/IEEE International Symposium on Integrated Network Management (IM); IEEE, May, 2021; pp 872–877.
21. Hopwood, S. How many Mobile Apps are Actually Used? June 22, 2017. https://www.apptentive.com/blog/2017/06/22/how-many-mobileapps-are-actually-used/
22. Ikram, M.; Kaafar, M. A. In *A First Look at Mobile ad-Blocking Apps*, 2017 IEEE 16th International Symposium on Network Computing and Applications (NCA); IEEE, Oct 2017; pp 1–8. DOI: https://doi.org/10.1145/2987443.2987471
23. Ikram, M.; Vallina-Rodriguez, N.; Seneviratne, S.; Kaafar, M. A.; Paxson, V. In *An Analysis of the Privacy and Security Risks of Android vpn Permission-Enabled Apps*, Proceedings of the 2016 Internet Measurement Conference; Nov 2016; pp 349–364. DOI: https://doi.org/10.1145/2987443.2987471
24. Johnson, V. L.; Kiser, A.; Washington, R.; Torres, R. Limitations to the Rapid Adoption of M-payment Services: Understanding the Impact of Privacy Risk on M-payment Services. *Comput. Hum. Behav.* **2018**, *79*, 111–122. https://doi.org/10.1016/j.chb.2017.10.035.
25. Johnston, A. C.; Warkentin, M. Fear Appeals and Information Security Behaviors: An Empirical Study. *MIS Q.* **2010**, *34* (3), 549–566. https://doi.org/10.2307/ 25750691
26. Kaptchuk, G.; Goldstein, D. G.; Hargittai, E.; Hofman, J.; Redmiles, E. M. How Good is Good Enough for COVID19 Apps? The Influence of Benefits, Accuracy, and Privacy on Willingness to Adopt, 2020. arXiv preprint arXiv:2005.04343.
27. Kang, J.; Kim, D.; Kim, H.; Huh, J. H. In *Analyzing Unnecessary Permissions Requested by Android Apps Based on Users' Opinions*, International Workshop on Information Security Applications; Springer, Cham; Aug 2014; pp 68–79.
28. Kim, D. J. Self-perception-Based Versus Transference-Based Trust Determinants in Computer-Mediated Transactions: A Cross-Cultural Comparison Study. *J. Manag. Inf. Syst.* **2008**, *24* (4), 13–45. https://doi.org/10.2753/mis0742- 1222240401
29. Krishnamurthy, B.; Wills, C. E. In *On the Leakage of Personally Identifiable Information Via Online Social Networks*, Proceedings of the 2nd ACM Workshop on Online Social Networks; Aug 2009; pp 7–12. DOI: https://doi.org/10.1145/1592665.1592668
30. Kumar, A. Risk of Mobile Threats and Privacy Concerns Grow; June 3, 2016. CSO Online. https://www.csoonline.com/article/3078815/security/risk-ofmobile-threats-and-privacy-concerns-grow.html.
31. Liccardi, I.; Pato, J.; Weitzner, D. J. Improving User Choice through Better Mobile Apps Transparency and Permissions Analysis. *J. Priv. Confidentiality* **2014**, *5* (2). DOI: https://journalprivacyconfidentiality.org/index.php/jpc/article/view/630
32. Levenson, H. 7 Common Reasons users are Abandoning your App; Aug 2, 2016 [Online]. Web Analytics World, https://www.webanalyticsworld.net/2016/08/why-usersare-abandoning-your-mobile-app.html.

33. Pagliery, J. Hackers are Draining Bank Accounts Via the Starbucks App.; May 14, 2015; CNN Business. https://money.cnn.com/2015/05/13/technology/ hackers-starbucks-app/index.html.

34. Pavlou, P. A.; Liang, H.; Xue, Y. Understanding and Mitigating Uncertainty in Online Exchange Relationships: A Principal-agent Perspective. *MIS Q.* **2007,** *31* (1), 105–136. https://doi.org/10.2307/25148783

35. Peddinti, S. T.; Bilogrevic, I.; Taft, N.; Pelikan, M.; Erlingsson, Ú.; Anthonysamy, P.; Hogben, G. In *Reducing Permission Requests in Mobile Apps*, Proceedings of the Internet Measurement Conference, Oct 2019; pp 259–266. DOI: https://doi.org/10.1145/3355369.3355584

36. Perez, S. Majority of U.S. Consumers Still Download Zero Apps per Month, Says comScore, 2017. https://techcrunch.com/2017/08/25/ majority-of-u-s?consumers-still-download-zero-apps-per-month-says-comscore/.

37. Posey, C.; Roberts, T. L.; Lowry, P. B.; Bennett, R. J.; Courtney, J. F. Insiders' Protection of Organizational Information Assets: Development of a Systematics-Based Taxonomy and Theory of Diversity for Protection-Motivated Behaviors. *Mis. Q.* **2013,** 1189–1210.

38. Posey, C.; Roberts, T. L.; Lowry, P. B.; Hightower, R. T. Bridging the Divide: A Qualitative Comparison of Information Security thought Patterns between Information Security Professionals and Ordinary Organizational Insiders. *Inf. Manag.* **2014,** *51* (5), 551–567. DOI: https://doi.org/10.1016/j.im.2014.03.009

39. Prasad, A.; Kotz, D. In *ENACT: Encounter-Based Architecture for Contact Tracing*, Proceedings of the 4th International on Workshop on Physical Analytics; June 2017; pp 37–42.

40. Reddy, E.; Kumar, S.; Rollings, N.; Chandra, R. Mobile Application for Dengue Fever Monitoring and Tracking Via GPS: Case Study for Fiji, 2015. arXiv preprint arXiv:1503.00814.

41. Reychav, I.; Beeri, R.; Balapour, A.; Raban, D. R.; Sabherwal, R.; Azuri, J. How Reliable are Self-assessments using Mobile Technology in Healthcare? The Effects of Technology Identity and Self-efficacy. *Comput. Hum. Behav.* **2019,** *91,* 52–61. DOI: https://doi.org/10.1016/j.chb.2018.09.024

42. Shah, M. H.; Peikari, H. R.; Yasin, N. M. The Determinants of Individuals' Perceived e-Security: Evidence from Malaysia. *Int. J. Inf. Manag.* **2014,** *34* (1), 48–57. DOI: https://doi.org/10.1016/j.ijinfomgt.2013.10.001.

43. Shahabi, C.; Fan, L.; Nocera, L.; Xiong, L.; Li, M. In *Privacy-Preserving Inference of Social Relationships from Location Data: A Vision Paper*, Proceedings of the 23rd SIGSPATIAL International Conference on Advances in Geographic Information Systems, Nov 2015; pp 1–4.

44. Sharma, T.; Bambenek, J. C.; Bashir, M. Preserving Privacy in Cyber-Physical-Social Systems: An Anonymity and Access Control Approach, 2020.

45. Sharma, T.; Dyer, H. A.; Bashir, M. Enabling User-Centered Privacy Controls for Mobile Applications: Covid-19 Perspective. *ACM Trans. Int. Technol.* **2021,** *21* (1), 1–24. DOI: https://doi.org/10.1145/3434777

46. Shaw, N.; Sergueeva, K. The Non-monetary Benefits of Mobile Commerce: Extending UTAUT2 with Perceived Value. *Int. J. Inf. Manag.* **2019,** *45,* 44–55. DOI: https://doi.org/10.1016/j.ijinfomgt.2018.10.024.

47. Shi, L.; Fu, J.; Guo, Z.; Ming, J. In *"Jekyll and Hyde" is Risky: Shared-Everything Threat Mitigation in Dual-Instance Apps*, Proceedings of the 17th Annual International Conference on Mobile Systems, Applications, and Services, June 2019; pp 222–235.

48. Simko, L.; Calo, R.; Roesner, F.; Kohno, T. COVID-19 Contact Tracing and Privacy: Studying Opinion and Preferences, 2020. arXiv preprint arXiv:2005.06056.

49. Statista. Average Number of New Android App Releases per day from 3rd Quarter 2016 to 1st quarter 2018, 2018. https://www.statista.com/statistics/276703/android-app-releases-worldwide/

50. Troncoso, C.; et al. Decentralized Privacy-Preserving Proximity Tracing, 2020. arXiv preprint arXiv:2005.12273.

51. Van Kleek, M.; Liccardi, I.; Binns, R.; Zhao, J.; Weitzner, D. J.; Shadbolt, N. In *Better the Devil you Know: Exposing the Data Sharing Practices of Smartphone Apps*, Proceedings of the 2017 CHI Conference on Human Factors in Computing Systems, May 2017; pp 5208–5220. DOI: https://doi.org/10.1145/3025453.3025556

52. Yoneki, E.; Crowcroft, J. Epimap: Towards Quantifying Contact Networks for Understanding Epidemiology in Developing Countries. *Ad Hoc Netw.* **2014,** *13,* 83–93.

53. Zuo, C.; Lin, Z.; Zhang, Y. In *Why does your Data Leak? Uncovering the Data Leakage in Cloud from Mobile Apps*, 2019 IEEE Symposium on Security and Privacy (SP); IEEE, May 2019; pp 1296–1310.

PART II
Cybersecurity for Cloud Intelligent Systems

CHAPTER 5

CLOUD AND EDGE COMPUTING SECURITY USING ARTIFICIAL INTELLIGENCE AND SOFT COMPUTING TECHNIQUES

N. S. GOWRI GANESH, R. ROOPA CHANDRIKA, and
A. MUMMOORTHY

MallaReddy College of Engineering and Technology, Hyderabad

ABSTRACT

Cloud computing extends its capabilities to the Internet of Things (IoT) devices at the endpoint by handling and processing user-centric data, apart from providing services for infrastructure, platform, and software. Edge computing is the technology extended to cloud computing for reducing the load in the cloud by processing the data at the endpoint of the networks. In Edge computing, cloudlets are placed at the edge of the network or in proximity to the end users/devices/applications. The major benefits are a reduction in bandwidth demands and roundtrip latency between end-user device and resource in the cloud. There are many assets such as routers, relays, and sensors participating in these environments. For cloud service providers, governance for security is a major concern to tackle threats and vulnerability. Artificial intelligence is an emerging field that has showcased its use in various applications like autonomous cars, speech recognition, automatic planning, scheduling, etc. Artificial intelligence (AI) applied to edge computing/fog computing shortly called Edge AI nowadays gains capabilities and features to provide decision-making to establish smart

The Fusion of Artificial Intelligence and Soft Computing Techniques for Cybersecurity.
M. A. Jabbar, Sanju Tiwari, Subhendu Kumar Pani, & Stephen Huang (Eds.)

devices connected to the cloud. These smart devices adapt to the environment dynamically by making intelligent decisions. The autonomous vehicle that uses Edge AI could adapt to the different environment conditions and the behaviors of drivers dynamically. Similarly, AI in edge computing also can be applied to control the movement of data within the cloud, IoT devices, and also users dynamically by making intelligent decisions. This feature could establish the security in the cloud and edge computing. This chapter deals about the security of edge computing and its associated environment by the application of AI and soft computing techniques.

5.1   INTRODUCTION

Cloud computing is the preferred environment for the deployment of software applications by the IT managers as it provides the benefits of infrastructure, a platform for development, and software in the form of services that are comfortable and easy to use. The users can utilize in their own pace of requirement of consumption by renting out the computing resources. The major benefits of cloud computing are elasticity, resource pooling, cost-effectiveness, adaptability, and measured services. Internet of Things (IoT) are devices having computational and communication capabilities which are distributed in large numbers with less processing and storage capacities that are seamlessly connected to the Internet. These limitations are balanced when IoT is integrated with cloud computing which provides the large storage capacity to store the data and computing power for data processing. However, all the data generated the IoT devices are not useful for processing. To process the required and desired data, edge computing bridges the appropriate devices at the endpoint and cloud computing. In Edge computing, devices are placed at the closeness to the instances of the event and sources of data. The locations of the devices are decided based on the decision for the immediate processing of the data which will offload the burden of the cloud infrastructure. The ideology of fog computing is closer to that Edge computing as the former deals with the software and platform as a service (SaaS/PaaS) for enterprise architecture above the Edge computing layer, whereas the latter deals with the services-related infrastructure ability (IaaS). Fog computing is a concept that is being in development in parallel to the Edge computing.

Soft computing approaches generally deal with approximate models comprised of approximate reasoning and approximate modeling. The soft computing methods such as fuzzy logic, neural networks, genetic algorithms, and expert systems are used to solve complex problems like speech recognition and pattern recognition. In the Edge/Fog systems, the soft computing approaches are confined to artificial intelligence (AI) and machine learning (ML) algorithms implemented in the cloud and Edge devices.

Context-aware applications are capable of obtaining, deciphering, and utilizing context information and complying with the current context. Intelligent agents in AI perform this perception of environment with the help of sensors and act accordingly with the support of actuators. These intelligent agents take decision like human agents with the help of necessary devices and can learn from the previous experience. The context-aware applications can be anything like speech recognition, workout tracking, and user activity recognition. Edge computing acts as the aid for the stationing of intelligent agents in proximity to the location of data source, which can collect information and acknowledge this flow of data. The edge devices are furnished with sensing devices that can track the environment. The intelligent agent housed in the edge devices can act independently or in combination with the other agent in the same devices or other devices to produce a meaningful action. When multiple autonomous agents work in tandem with each other to solve a comprehensive problem they are termed as multi-agent system.

Security at the IoT/edge nodes is the utilization of security practices at the nodes of the network outside the core network. Generally, network attacks are the activity that breaches the privacy rules, changes processed data, modify a certain instance to make all or any of the network devices/software inaccessible and/or unreliable compromising on confidentiality, data integrity, and exposes denial of service (DOS) attack, respectively. The software using AI for security learns about the past behavior and makes quick decisions for appropriate action if new/unknown behaviors/information is submitted. ML algorithms and deep learning (DL) techniques play a vital role in the development of AI. These AI modules also provide logical inferences on the incomplete data and different alternate solutions to a known problem to assist the security team to select the way for tackling the security issues. The implementations of AI/ML at the edge devices referred to as Edge AI have the potential to enhance the data privacy and impede the challenges that arise due to threats at multiple endpoints.

5.2   IOT AND CLOUD COMPUTING

The IoT devices have sensors and mini-processors which can process the sensors data and can be transmitted to the cloud via internet for further processing and storage. It interacts with the billions of devices that exchange data within the things or real-world objects. The objective of the IoT is to empower things/objects to be linked to each other at any point of time pervasively using any route/network and any service. Generally, it uses constrained application protocol (CoAP), message queue telemetry transport protocol (MQTT), extensible messaging and presence protocol (XMPP), advanced message queuing protocol (AMQP), and data distribution service (DDS). The data generated by the IoT devices are processed at the cloud computers. The five layers of IoT[28] architecture are perception layer, transport layer, processing layer, application layer, and business layer as shown in Fig. 5.1.

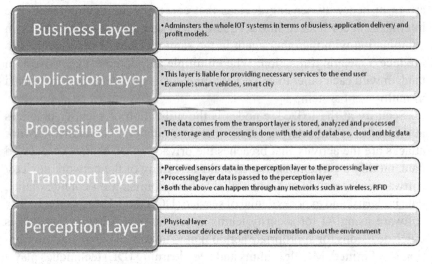

FIGURE 5.1   Five layers of IoT.

The IoT devices are categorized[6] as class 0 or low-end devices, class 1 or middle-end devices, and class 2 or high-end IoT devices. The examples for class 0 are openmote, waspmote, sensor, and actuator; for class 1 are Arduino, Netduino, and Gateways; for class 2 are Rasperry pie and

Beagleboard. The operating systems available for IoT devices are low-end Linux-based like ARM Mbed and LiteOS and high-end Linux-based like Raspbian and Android Things. Non-Linux OS like TinyOS and free RTOS are also available.

Cloud computing is a great innovation in distributed technology in pursuit to the development of virtualization techniques and improvement in web services to enable any application in the form of services. It is a boon to startups that in the short period large numbers of new ventures are being coming to the Industry. It reduced the infrastructure set up cost by providing computing, storage, platform, development, and software in the form of services such as Infrastructure as a Service (IaaS), Platform as a Service (PaaS), and Software as a Service (SaaS). The organizations can create their own private cloud or attach themselves to the public cloud or have hybrid cloud in combination of both their organization's private and public cloud. Due to the interesting and important property of elasticity, the services can be utilized by the organization attaching to the cloud to consume huge computing power and large storage space during the peak demand and can be diminished according to their requirement. Even though cloud and IoT evolved independently, IoT gets assistance from cloud[8] which offers many services as described in Fig. 5.2. There are many IoT cloud platforms[26] discussed for the purpose of application development, device management, heterogeneity management, and monitoring management such as Echelon, Carriots, Oracle IoT cloud, and Exosite.

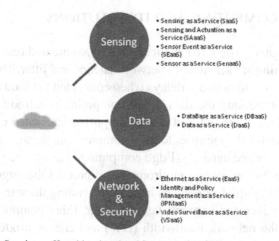

FIGURE 5.2   Services offered by the cloud integrated to IoT.

The integration of IoT and cloud provides a lot of benefits such as capturing of real-time data, decision-making, analysis of data, and management of devices. There are many applications of IoT combined with cloud. They are smart health management, smart agriculture, smart logistics, accidental management, smart city, etc., as indicated in Figure 5.3 .

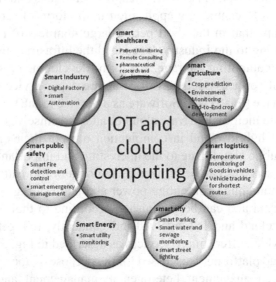

FIGURE 5.3   Applications of IoT and cloud computing.

5.3   EDGE COMPUTING AND ITS SOLUTIONS

Applications that involve IoT devices such as augmented reality, e-health-care, and gaming are sensitive to network latency and jitter. These applications are sensitive to network delays when connected to cloud even though it assists in processing the data. Edge computing is introduced to solve the above issues in the IoT and cloud computing integrated environment. Edge devices have the facilities to have computation, storage, and communication with limited capacity. Edge computing is to use the edge devices located nearer to the context environment to process the large amount of data generated by the IoT devices instead of sending them to cloud infrastructure to reduce the network latency, jitter. Edge computing in addition to save the network bandwidth (BW) and energy produces quicker response with better quality.

The core concept of cloud computing is virtualization. The motive of virtualization is that the individual robust set of hardware needs to achieve the function of several sets of hardware. The same virtualization technology influences the capacities of Edge devices as well. There are a few variations in the type of virtualization that depends on the system that participates in the Edge computing environment. Multitenancy, scaling with elastic approach and tight resource management are the characteristics that best describes hardware virtualization. Wireless sensor network virtualization[17] allows multiple wsn applications/services to exist side-by-side against a single application which leads to the development of innovative wsn applications. In OS-level virtualization, Linux containers (LXC) are adopted to execute the services at the edge devices. Further, Docker-based containers are applied in the Edge devices to improve performance and interoperability among the Edge nodes. Edge hardware[10] is basically categorized as Edge devices and Edge/Fog Servers. There are three types of Edge devices that are categorized into devices that support virtualization and vice versa. (1) Constrained devices are any IoT devices which do not support virtualization. (2) Embedded computer controllers or single-board computers like Intel NUC or Raspberry Pi supports both software and hardware virtualization. (3) Mobile devices such as smartphones act at the edge of the networks support to full virtualization. Edge/Fog servers offer services to these Edge devices.

Three edge computing solutions for IoT are namely cloudlet, mobile edge computing (MEC), and Fog computing.

Cloudlets are small scale of clouds located nearer to the Edge devices. The cloudlet is a high-end computer or computers connected in clusters that are connected to the internet. The general arrangement architecture for cloudlet is three-tier architecture having one end the mobile device layer, middle cloudlet layer, and the other end is the cloud layer. The cloudlets use virtual machines (VM) technology for the purpose of rapid provisioning.

Mobile edge computing[24] standardized by the European Telecommunications Standards Institute (ETSI) offers the capabilities of cloud computing environment and IT services management at the edge of the mobile network and in the vicinity to mobile users. The purpose is minimizing latency, better service delivery, and mobile subscriber experience. The concept of MEC also works based on the virtualization infrastructure where the MEC applications live. ETSI standards had issued the reference architecture that describes the working principle of MEC.

The notion of the existence of Fog computing is that the enablement of the computation available in the cloud to be also made available in the IoT Sensors and Actuators. So Fog computing is defined by openconsortium as openfog reference architecture that distributes resources and services of computing, storage, control, and networking anywhere in continuity from the cloud to things. It describes three views namely, software view for application services, application support and software backplane, system view to describe about hardware virtualization and Node view to deal about the protocol abstraction layer and sensors, actuators, and control. The main factors considered in the reference architecture are referred to as key pillars. They are security, scalability, openness, autonomy, RAS, agility, hierarchy, and programmability.

The Edge computing resource management approaches are many due to the variety of devices, usage of assortment of different networks, and ubiquitous in nature. Three resource management levels are categorized for this purpose; they are back-end infrastructure, edge systems, and front-end smart devices. The resource management approaches for (1) back-end infrastructure are task placement and hybrid provisioning. (2) Edge-system levels are adaptive provisioning, load coordination, service migration, data preprocessing, computation offloading, and distributed caching. (3) Front-end smart device is resource pooling/sharing.

5.4 SECURITY IN EDGE COMPUTING

The objective of managing the security in a system or system that is offering services is to manage the integrity, confidentiality, and availability of the data/services. The security is dealt (Fig. 5.4) in sequence of computer security, data center security, cloud computing security, and Edge computing security.

FIGURE 5.4   Security.

The exposure of a system to attacks is possible for various reasons. The external entity attempt for getting entry or spoil the system either to gain from the resources available in the system or to destroy the available data. Such kinds of external entities are vulnerable to the system. The security architecture in every system needs to take two primary steps: (1) Prevention by early identification: to expect these external vulnerable entities and actions to prevent their entry to be taken at the early stage. (2) Counter the attacks: actions that need to be taken when vulnerable entity attacks the system. Risk management supports the first primary activity which identifies the risks and assists to prevent the same. Security algorithms such as symmetric, asymmetric and NP-hard algorithms support the system to counter the attacks. Generally, attacks can be classified into the external attacks and internal attacks. System attack is a type of internal attack where in which the files are deleted or modified, password will be stolen for an entry without the knowledge of the owner. Network attack and web attack are external attacks. These attacks try to alter the data exchanged in the network or monitor the flow of data or trap the data to make use of it. Backdoor method, DOS attack, Eavesdropping, Phishing, Privilege escalation, Tampering, Malware, IP spoofing, ICMP flooding, and Traffic analyses are some of the examples of attacks. To prevent attacks, user account controls, cryptography, firewalls, and network intrusion detection (NID) are few among the security techniques used.

An organization running web applications like e-commerce, generally will be having a secured data center which acts as a centralized or distributed server. A data center in an organization has IT infrastructure comprising of network of computers, storage systems for the purpose of organizing, processing, and storing huge amount of data. Security at the data center level refers to the set of practices and implementation of physical and virtual technologies to prevent the data center from possible internal and external attacks. Data center ranges from simple setup to higher tier facilities (Tier 3 or Tier 4) which are larger and more complex environments. Generally, security for data center is classified as physical security and software security. Physical security deals with the location of the datacenter, access to the datacenter, cooling arrangements, etc. Creating secure zones for having layers of network security, installing necessary software for detecting and monitoring vulnerabilities, and tools like security information and management tools to have real-time view of the data center security posture are some of the Software security arrangements in the datacenter. Normally, practitioners will implement the data center security which refers to the precautionary measures as defined in the various standards. The notable data security standards are ISO/IEC 27001:2005 and 27001:2013 Information Security Management System Standards, NIST 800-53 – Security and Privacy Controls for Federal Information Systems and Organizations. The methodology[21] to inspect and verify the security aspects from different dimensions are required which may include security control analysis, maturity evaluation, and result analysis. The information technology management systems[2] are developed in the data center that can support to find, maintain, and curtail information security threats.

Cloud computing is highly acknowledged by organizations that are required to host the infrastructure and software setup at lightning speed and less cost per usage. Due to its popularity, the cloud is increasingly used by many in spite of its security challenges. Among the various deployment models such as private cloud, public cloud, and hybrid cloud, the exposure to the attacks is more for the public cloud and less prone for the private cloud.

There are various security concerns with respect to the cloud delivery models[30] for SaaS, PaaS, and IaaS. SaaS application vendors provide the application in the cloud computing environment for the enterprises readily without or with the little configuration without the need of the installation in their premises. This is a great attraction for the software customers so that if they are interested in the application they can continue with that or else they drop it and they are expected to pay only for the services they consumed as per the service level agreements. The data in the SaaS environment are the focus of subject regarding the security. As the SaaS is offered for multiple enterprises, the management of data by the SaaS providers of various customers will be stored in the same storage location. As per the high availability policy, the data of the SaaS applications may be stored in multiple locations. PaaS allows the developers to build the applications in the cloud without the need to set up the development environment in their system rapidly. The main concept of PaaS relies on the web services. The problem is that when the service of PaaS fails suddenly what will happen to the applications and the data running on it will become a major issue. Ws-security at the cloud security gateways needs to be strengthened for the protection of data. IaaS offers the developers to set up the necessary infrastructure quickly. When the data get stored and moved to other devices there is a possibility that these data can move across the devices of the intruders. In addition to strong encryption and protocols, strong policy for cloud environment is essential to keep the data secured.

All the above models depend on the virtualization technology implemented in cloud computing. In the virtualization, there will be an administrator that controls host and guest operating systems. The features of virtualization software are sharing, aggregation, emulation, and isolation. Isolation refers to the characteristics that a virtual machine (VM) is completely separated from its host system and other VMs such that it acts independently. Even though one VM fails it will not affect the other one or the host system. Though the virtualization softwares are required to have the property of isolation, it is yet to be implemented in full considering such security issues. The virtual machine monitor (VMM) does not offer perfect isolation. Vulnerabilities are detected such that users with the wrong intent can use malicious code in the virtualized guest environment can gain access to the host environment.

Edge computing is the environment to process the data with less network latency and better quality output, where data are the core part that is vulnerable to external attacks. As the Edge nodes are available at the close proximity to the environment where data acquisition and computing are required, it obviously lacks the physical security that is available in the secured data center.

Security in Edge computing is dealt with in terms of securing the data and privacy protection[9] as follows:

(1) Secure data sharing:
   Secure data sharing main purpose is to ensure confidentiality which is implemented with encryption techniques like attribute-based encryption, homomorphic encryption, etc.
(2) Integrity of the outsourced data in the cloud:
   Integrity of the data stored in the edge or cloud data center is established by dynamic and batch audit.
(3) Search on the cipher text:
   This feature[19] enables the query of cipher text data and their retrieval. Searchable keywords and its related data are encrypted. As IoT devices do not have the capability to perform encryption/decryption, the process will be administered by EdgeServer/FogServer/Cloud.

(4) Authentication in single domain, multiple domain, and in handover. To access computing and storage services, each entity must pass authentication. Same is the case for multiple domains. For example, Microsoft Edge[11] allows the entities to authorize access to the resources with the help of single windows integrated authentication (WIA) in single domain and seamless single sign on (SSO) access in multiple domains. Table 5.1 describes some of the authentication standards available for cloud and Edge computing.

TABLE 5.1   Authentication Standards.

OAuth2.0	It is security standard designed for authorization that is to grant access from one to another application.
OpenID	Open standard which allows users to be authenticated by relying parties. Auth0, Okta are some of the authentication provider
Identity Provider(IdP)	It is the entity that creates and manages identity information for principles and provides authentication services for applications created by relying parties, e.g., Azure AD, Google, Salesforce
JSON Web Token (JWT)	It creates access tokens with JSON based open standard

(5) Data, identity, and location privacy.
   The fetching, processing of data in edge computing happens across various devices such as IoT, Edge Node, Edge Server, and cloud. The data and identity of the users need to be available securely across all the environments. The location of usage of mobile in the edge varies depending upon the movement of the user. The information about this needs to be protected and there is a possible use of this information in malicious intent which may result in harmful consequences. Chaff control strategies[14] are developed to control location privacy.

(6) Access control for system security and privacy of the user
   To protect system security and user privacy, the access control system is the technique used in Edge systems. There are many advantages of access control mechanisms[3] which helps (1) only authorized users to access the sensitive data and others with wrong intent to be curbed in accessing these data. (2) to avoid the hackers to access the physical part of the system. (3) to secure the cyber infrastructure and protect from attacks. While designing access control mechanisms in Edge/Fog following requirements[38] need to be considered:

(1) To gain access to the computing and storage of cloud/Edge access control policies need to be in place. (2) Access controls are required to both the directions of access from Fog/Edge to cloud and vice-versa. (3) Access control mechanisms are to be adapted to virtual machines (VM) to prohibit any attacks. There are various models available for designing access control mechanisms. They are role-based access control (RBAC) model, discretionary access control (DAC) model, mandatory access control (MAC) model, attribute-based access control (ABAC) model, usage-control-based access control (UCON) model, reference monitoring access control (RMAC), and proxy re-encryption (PRE) model. In ABAC model, attribute-based encryption (ABE) is used by the data owner who decides and frame policies on who can access which data. In spite of some of the challenges like managing access policies, ABE-based access model is considered to the suitable technique for the edge computing environment by many researchers.

Edge computing environment faces many challenges in ensuring security. Smart devices have some constraints typically in its computational, power, and storage capabilities. Further these devices are heterogeneous in nature. So the security solutions such as installation of antivirus software on these devices and conventional asymmetric key encryptions algorithms cannot be used.

5.5 SOFT COMPUTING AND ARTIFICIAL INTELLIGENCE

Zadeh[27] had classified machine intelligence into hard computing based on artificial intelligence and soft computing based on computational intelligence.

Soft computing[16] accords solutions to complex real-world problems which is based on approximate models and is tolerant of approximations, imprecision, partial truth, and uncertainty to accomplish tractability, low solution cost, and robustness. It has the characteristics of qualitative, dispositionality, and approximation. Fuzzy logic, genetic algorithms, expert systems, and artificial neural networks (ANN) are some of the soft computing methods.

Artificial intelligence[34] is coined in the year 1956 after the World War II in the 2-month workshop held at Dartmouth. It deals with perception, understanding, predication, and manipulating which can build intelligent machines. An AI system can be described essentially as a knowledge-based agent comprising of the components—perceptors and actuators to sense and act back with the external environment. Also the internal components would be a problem-solving engine and learning engine using knowledge base system which is useful to draw inferences, solve problems, and decide on actions upon receiving inputs from the sensors or other perceiving devices.

Early methods of AI when applied to real-world problems lead to failures. The researchers were able to understand that building an intelligent agent is very hard because the cognitive functions used for automation are not understood clearly in depth. As a result, in order to understand the concepts deeply, AI is split into knowledge representation, learning, probabilistic reasoning, neural networks, and many other areas.

5.6 MACHINE LEARNING TECHNIQUES

The literature[22] has focused that learning and adaptation are the most important aspects of intelligence to be modeled in a computer for a machine to mimic a human. Modeling is to train machines to teach and adapt to the environment from the data it has acquired. Machines are taught to efficiently handle data to choose the actions to perform on the environment appropriately, like controlling a robot or making future predictions. The idea of ML is that the actions performed need not be explicitly programmed in a dynamically changing environment but learn from its history of sequences. ML techniques and ideas are exploited from biological objects, mathematics, statistics, and neuroscience.

The techniques try to automatically identify patterns of data to predict the future outcomes. ML algorithms are categorized based on the outcomes of the algorithm.

The learning algorithms[4] are :

- Supervised learning
- Unsupervised learning
- Semi-supervised learning
- Reinforcement learning
- Learning to learn

Supervised or predictive learning is to map the inputs to desired outputs given a labeled set of data. It is also called as learning from exemplars. The labeled set of data is the training set that consists of input features or attributes or covariates. The desired outcomes are the output variables called as a response or target variables. There are two types of supervised learning namely, classification and regression. In classification problems, the target variables are categorical type, and in regression, the variables are continuous values. The performance of a supervised algorithm is how well it generalizes, in cases with an unexpected input; the algorithm will give a desired sensible output. The different algorithm types are linear classifiers, k-means clustering, Bayes classification, neural networks, and decision trees.

Unsupervised learning is also called as descriptive learning. The algorithm tries to discover new patterns from the given input set. This approach is also called as knowledge discovery. Compared to the supervised learning, unsupervised learning does not have a well-defined problem. The algorithm fits into decision framework problems.

Semi-supervised learning is between supervised and unsupervised learning. This approach analyzes a large amount of unlabeled datasets in combination with a few labeled dataset.

Reinforcement learning is an approach that learns its actions from trial and error interactions with the environment. The agent's action set an impact on the environment and in return receives a feedback from the environment that helps the learning algorithm to improvise its behavior.

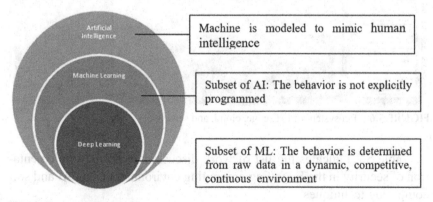

FIGURE 5.5   Subset of AI.

Learning to learn techniques is based on previous experience. A recent study involves DL that extracts raw data from sensor/IoT devices. This type of learning is challenging because the environment is complex and the data received may not be accurate. Figure 5.5 shows the relationship between AI and its subsets.

5.7   APPLICATIONS OF AI AND SOFT COMPUTING TECHNIQUES TO IMPLEMENT SECURITY IN CLOUD AND EDGE COMPUTING

Security in the edge computing environment is essential and need the assistance of other technologies. Due to the low computing power and less storage characteristics of IoT and Edge devices, the data need to be sent to edge servers/Fog servers/cloud computing. The data may be intruded by the hackers. Application of the encryption algorithms at the IoT devices is of less feasibility as it requires large

computing power. Figure 5.6 illustrates a generic ecosystem that demonstrates the need of security in IoT, Edge, Cloud, and application layers. Many researchers' implemented security using soft computing and AI, particularly ML algorithms are applied by at various levels of the edge computing environment.

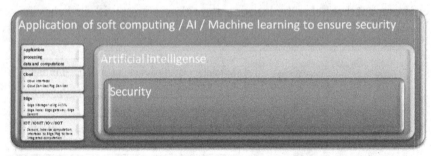

FIGURE 5.6   Ecosystem of IoT, edge, cloud, and security.

Table 5.2 indicates the list of work discussed about the implementation of security in the Edge/Fog computing environment using AI and soft computing techniques.

TABLE 5.2 List of Implementation of Security in Edge Computing using AI/Soft Computing.

Type of soft computing/ AI	Technique	Security aspect
Soft computing	Fuzziness-based semi supervised learning	Ensemble learning	Network intrusion detection
	Fuzzy inference system	MEC with security service chaining	To find the proper order of required security functions
	Soft hesitant fuzzy rough set	Multi-criteria decision-making	To handle several dynamically varying security services with the mobile user's requirements
Artificial intelligence	Deep learning	Fog learn	Data privacy
	AI agent	Self-mediating agent	Trustless protection framework
	CNN model	DeepNFV	Network traffic analysis
	CNN model	IoT guard	Real-time security management system
	Deep neural network	CryptoNets	Encryption of data
	Decision tree approach	Cyber attacks on robots operation	Intrusion detection
	Deep learning-based binary decision diagram multi-layer extreme learning	Finger-vein recognition systems	Data privacy preservation of finger-vein biometric data
	J48, Byes Net, RandomForest, Hoeffding, support vector machine (SVM) and deep learning	AI-based reaction agent	Anomaly-based intrusion detection system
	Support vector machines (SVM), clustering, bagging, and deep learning (DL) algorithms such as convolution neural network (CNN) and long short-term memory (LSTM) applied on blockchain-based data	Data analytics on the block chain-based security problems	Intrusion detection system

Note: The "Type of soft computing/ AI" column value applies to the grouped rows as shown.

TABLE 5.2   *(Continued)*

Type of soft computing/ AI	Technique	Security aspect
MapReduce-based system, the graph model-based abstraction system, and the parameter server system	Distributed machine learning-oriented data integrity verification scheme (DML-DIV)	Data integrity verification
Deep reinforcement learning approach	Deep Q-network	Data snooping and alteration reduction
Neural network	Para-cloud in edge nodes	Data privacy and protection
AI- Local Binary Pattern – Face Recognition	AI edge and cloud computing (AE-FRS)	Human identification system

Despite the fact that the cloud-based robotic system[12] has provided services in a variety of industries, its data protection is constantly threatened, and the network intrusion detection system (NIDS) is regarded as a critical component to ensure its security. Many ML techniques have been used to create a more intelligent NIDS in recent years. The majority of NIDSs focused on ML and AI are either supervised or unsupervised. However, supervised learning for NIDS is heavily reliant on labeled data. This flaw makes it more difficult to identify the most recent attack patterns. Unsupervised learning for NIDS, on the other hand, often fails to produce adequate results. As a result, this paper presents a new fuzziness-based semi-supervised learning approach for network intrusion detection on a cloud-based robotic system through ensemble learning, which can resolve the aforementioned issues. First, we build an ensemble system trained on labeled data due to ensemble learning's strong generalization capability. Furthermore, a fuzziness-based approach for data processing is used to better use the unlabeled data. In this way, the dataset's noisy and redundant examples are omitted. Finally, it integrates both supervised and unsupervised pieces using the same ensemble approach. The proposed solution is checked on the NSL-KDD dataset to check the efficacy and robustness of the NIDS.

Starting with static vulnerability analysis, framework[36] is built that implements zero confidence. "Zero trust" is a model, and there are three main concepts that must be followed in order for any network in modern computing to be safe across boundaries. These principles are: (1) ensure that all services, regardless of location, are accessed securely; (2) use a strict least privilege approach; and (3) inspect and document all traffic. The AI's position as a selfless mediating agent is investigated in order to address some issues in the implementation of a trustless protection framework, as well as the challenges that this poses. Time, expense, resource consumption, and degrees of freedom in the logical subsystem will be the most important control factors for a self-regulating trustless agent. These are important considerations for any mediator. In order to account for consumption trends and underlying dynamics, an AI agent may be a very useful regulatory agent.

One of the ML algorithms, such as CNNs form the next generation of cloud security[31] because CNN can provide automated and sensitive approaches to improving cloud security. ML may include solutions that implement comprehensive algorithms for stable enterprise data across all cloud applications, rather than concentrating solely on detecting and identifying sensitive data patterns.

A novel distributed DL scheme[1] is recommended for cyber-attack detection in fog-to-things computing. The master fog node sets the DL parameters and sends them to the worker nodes, but training data and hyper-parameter optimization are done locally on the worker nodes. On each worker node, parameter updates are made using optimizers and DL gradients, which are then aggregated on the master fog node. The master node receives asynchronously the updates of parameters from each node when the workers are ready to submit them. When a worker has several threads to run, it averages each thread's update and sends it back to the master node. The model is important for a number of reasons: worker intrusion detection systems (IDSs) detect malicious events locally, while the master IDS is responsible for updating the parameters using gradient descent training attack detection systems employing DL models on distributed IoT networks backed by fog nodes could improve the accuracy and efficiency of cyber-attack detection by sharing the parameter.

The main goal of this paper[31] is to use network traffic data to recognize and detect suspicious activities. For experimenting and performance evaluation on the CNN-MSVM process, two separate sets of data were used: TOR and UNSW and ISOT. All of the input data (network traffic data) are fed into the CNN model's input layer during the experiment. The main goal of this project is to develop a better learning method for data analytics in the context of cloud protection. It is also focused on putting a lot of investigative work into practice by combining CNN with the MSVM approach. The extended supervised ML methods are highly suitable and applicable in real-time cloud applications, according to the experimental results and output comparison.

DeepNFV[20], a lightweight network function virtualization (NFV) architecture based on the Docker container running on the network edge that combines state-of-the-art DL models with NFV containers to solve some complex issues, such as network traffic classification and connection analysis. Since this container is part of an NFV chain that the system chooses for a particular user, its input may be the user device or the output from other NFV containers. To process raw network traffic, the first step is to divide it into discrete units, or network packets. The packet capture (pcap) file, which is the universal file format in traffic analysis and can be used with the libpcap library in Unix-like systems, is used to record these units. Second, the packet headers should be changed to eliminate any unwanted or interfering data. The created images will be labeled with different labels during the training process of the proposed deep model, according to the specific protocols of their raw packets. The videos, as well as the labels that go with them, will then be imported into the CNN model for fine-tuning. The image generation module will directly pass the created images to the well-trained CNN model, which can provide deep insight into the pattern information, using the stochastic gradient descent (SGD) method.

For an intelligent, resource-efficient, and real-time security management system, develop[32] and implement IoT-guard, a distributed IoT architecture. In a smart home setting, the device, which consists of edge–fog computational layers, can help in crime prevention and predict crime events. The IoT guard can identify and confirm crime incidents in real time, using AI and an event-driven approach to send crime data to protective services and police units, allowing for rapid intervention while conserving resources including electricity, BW, memory, and CPU use. In this report, we construct a prototype of an IoT-guard laboratory testbed and assess its efficacy for real-time protection applications. A CNN model operating on a fog node identifies and saves images labeled with the names of the crime objects with the highest likelihood. After that, the fog node compiles and sends crime info.

Although an edge server[23] can speed up DNN processing, it is not always sufficient for edge devices to run DNNs on servers. There are three offloading scenarios: (1) partial offloading of partitioned DNN, (2) hierarchical architectures (offloading is done through a mix of edge computers, edge servers, and cloud), and (3) distributed computing approaches—the DNN computation is distributed across multiple peer devices. To compute the inference with a high degree of secrecy, cryptographic techniques can be used. Secure computation aims to ensure that an inference result is sent to the end system without the edge server knowing anything about the DNN model, and vice versa. Homomorphic encryption, which encrypts exchanged data and allows computation on the encrypted data, is one type of protected computation used in CryptoNets. The DNN is translated into CryptoNets, which are low-degree polynomials that approximate the common activation functions and operations in DNN, ensuring homomorphic encryption. The compute times of homomorphic encryption, on the other hand, appear to be a bottleneck. Another method for stable computation is multiparty computation. Multiple machines collaborate and interact in multiple rounds to jointly compute a result in safe multiparty computation.

Cyber-physical device intrusion detection is a relatively recent field of research. It has been studied to some degree in the context of industrial control systems, but it is still in its infancy in the context of mobile cyber-physical systems like robotic vehicles. It is discovered in the experimentation[35] that different attacks have different effects on the robot's operation, including both its cyber (network, CPU, and disk data) and physical behavior (speed, vibration, and power consumption). This provides an advantage since combining the two types of features will increase the accuracy of attack detection. This has been checked with a decision tree-based intrusion detection approach.

A privacy-preserving edge finger-vein biometric system[37] based on DL has been developed. Raw finger-vein templates are protected by the binary decision diagram (BDD), built and incorporated with the function permutation technique to make the transformation non-invertible and the transformed template revocable in the proposed finger-vein recognition scheme, that is, binary decision diagram multi-layer extreme learning machine (BDD-ML-ELM). The proposed system improves on existing permutation-based cancelable biometrics and ML-based finger-vein recognition systems, both of which lack template security.

A new ML-based security architecture that automatically addresses the growing security concerns in the IoT domain is suggested[5] in a work. For threat mitigation, this architecture uses both software defined networking (SDN) and NFV enablers. This AI platform incorporates a monitoring agent and AI-based reaction agent, both of which use ML models to analyze network patterns and identify anomalies in IoT systems. To achieve its objectives, the architecture makes use of supervised learning, a distributed data mining method, and a neural network. The distribution of attacks using the data mining technique, in particular, is extremely effective at detecting attacks with high performance and low cost. The test is conducted with the experiment in a real smart building scenario using a one-class support vector machine (SVM) for the anomaly based IDS for IoT.

Machine learning[33] is being used to improve the security of blockchain-based smart applications. Blockchain technology's consensus process ensures that data are safe and legitimate. New security problems, such as majority attack and double-spending, arise as a result of blockchain. Data analytics on blockchain-based encrypted data is suggested to fix the above-mentioned issues. ML is a method of making specific decisions using a logical amount of data. Both technologies can be used together in a variety of smart applications which utilized the edge computing and cloud computing environment.

In distributed machine learning (DML) like MapReduce-based system, the graph model-based abstraction system, and the parameter server system, ensuring data integrity is critical. If network attackers forge, alter, or destroy data, the DML system's training model will be severely harmed, and the training results will be incorrect. To ensure the integrity of training data,[39] a DML-oriented data integrity verification scheme (DML-DIV) is suggested. To achieve data integrity verification, the provable data possession (PDP) sampling auditing algorithm is used. This allows the referred scheme to withstand forgery and tampering attacks. Second, in the TPA verification process, random number is generated, the blinding factor, and discrete logarithm problem (DLP) is used to establish proof and ensure privacy security. Finally, we use identity-based cryptography and two-step key generation technology to produce the public/private key pair for the data owner so that this scheme can solve the key escrow problem and lower the cost of certificate management.

With cloud storage and network service offloading at the edge of mobile cellular networks, MEC enables energy conservation. However, a critical problem impeding potential MEC growth is how to effectively handle various real-time evolving security functions. A fuzzy security service chaining approach[18] for MEC is proposed to address this problem. A new architecture, in particular, is being developed to decouple the necessary security functions from the physical resources. A security proxy based on this is devised to enable compatibility with conventional security functions. A fuzzy ranking system to determine the best order for the appropriate security functions.

Tasks are offloaded to edge servers by mobile apps in the MEC environment. These are vulnerable to external security threats (e.g., snooping and alteration). In order to address this problem, a security and cost-aware computation offloading (SCACO) strategy is proposed for mobile users in a MEC environment, with the aim of reducing total costs while meeting risk likelihood constraints. The costs could be the mobile device's energy consumption, processing delay, and task loss probability. The computation offloading problem is first formulated as a Markov decision mechanism (MDP). The optimal offloading policy[15] for the proposed problem is then extracted using the common deep reinforcement learning approach, deep Q-network (DQN).

EdgeSec,[29] an Edge layer security service to improve the security of IoT systems EdgeSec is made up of seven major components that work together to address unique security issues in IoT systems in a systematic manner.

The FogLearn[7] framework was used to assist and thus improve the cloud computing framework's capabilities. The data are processed by the Fog layer in FogLearn, and after processing, the data can be sent to the cloud layer for long-term storage and analysis. As a result, this architecture gives end-users more power for improved efficiency without adding computational overhead at the cloud layer. When the computational overhead at the cloud layer is very high and geospatial data sizes are growing, the built architecture has an advantage. As a result, the fog layer boosts overall performance by lowering latency and increasing throughput. When the data are processed locally, this system brought a privacy advantage.

A critical problem impeding the growth of FMEC is how to effectively manage many dynamically varying security resources while meeting the needs of mobile users. To address this problem, we wanted to implement a method for selecting an acceptable security service in FMEC based on the needs of mobile users. The problem of choosing the right security service with hesitant fuzzy data is a multi-criteria decision-making problem. The authors presented[25] a soft hesitant fuzzy rough collection (SHFRS) to solve multi-criteria decision-making problems in this paper. By fusing the hesitant fuzzy rough set theory with the hesitant fuzzy soft set, SHFRS is implemented as a groundbreaking extension of the hesitant fuzzy rough set theory.

Real-time video processing is a crucial component that enables a variety of applications that would otherwise be impossible to implement due to scalability issues. When analyzing real-time information, predictive models, specifically neural networks (NNs), are often used to reduce processing time. Using NNs, on the other hand, is computationally costly. By delegating parts of the job to edge nodes, edge computing aims to build systems that alleviate the pressure on the cloud. In order to reduce latency, such systems prioritize processing as much as possible on the edge node before delegating the load to the cloud. Furthermore, processing information at the edge improves data privacy and protection. This paper utilizes[13] the edge nodes even more by allowing them to collaborate as a para-cloud, reducing reliance on the primary computing cloud.

High transmission rate and large data volume are two characteristics of visual sensors. Face recognition sensors are commonly used in defense, healthcare, and other fields. This paper proposes a hybrid of edge-based AI and cloud computing for applications such as face recognition and protection that necessitate a large number of visual sensors, as well as image processing and analysis.

5.8 CONCLUSION

Edge computing environment provides lot of facilities for IoT devices. But due to the heterogeneous nature of its positioning varieties of threats are prevalent. Many researchers have dealt with the security issues and soft computing/AI has provided the solutions as per the discussion in this chapter. It is expected that lot many types of new IoT devices are yet to come which may come up with the more different types of threats.

KEYWORDS

- soft computing
- artificial intelligence
- edge computing
- fog computing
- IoT
- cloud computing

REFERENCES

1. Abebe, A.; Chilamkurti, N. Deep Learning: The Frontier for Distributed Attack Detection in Fog-to-Things Computing. *IEEE Commun. Mag.* **2018,** *56* (2), 169–75. DOI: 10.1109/MCOM.2018.1700332.

2. Dedy, A.; Suryanto, Y.; Ramli, K. In *On Developing Information Security Management System (ISMS) Framework for ISO 27001-Based Data Center*; 2018 International Workshop on Big Data and Information Security (IWBIS), 2018; pp 149–57. DOI: 10.1109/IWBIS.2018.8471700.

3. Alramadhan, M.; Sha, K. In *An Overview of Access Control Mechanisms for Internet of Things*, 2017 26th International Conference on Computer Communication and Networks (ICCCN), 2017; pp 1–6. DOI: 10.1109/ICCCN.2017.8038503.

4. Ayodele, T. O. Types of Machine Learning Algorithms, In *New Advances in Machine Learning*; IntechOpen, 2010. DOI: 10.5772/9385.

5. Bagaa, M.; Taleb, T.; Bernabe, J. B.; Skarmeta, A. A Machine Learning Security Framework for Iot Systems. *IEEE Access* **2020,** *8*, 114066–114077. DOI: 10.1109/ACCESS.2020.2996214.

6. Bansal, S.; Kumar, D. IoT Ecosystem: A Survey on Devices, Gateways, Operating Systems, Middleware and Communication. *Int. J. Wirel. Inf. Netw.* 2020, *27* (3), 340–364. DOI: 10.1007/s10776-020-00483-7.

7. Rabindra, B. K.; Priyadarshini, R.; Dubey, H.; Kumar, V.; Mankodiya, K. FogLearn: Leveraging Fog-Based Machine Learning for Smart System Big Data Analytics. *Int. J. Fog Comput. (IJFC)* 1 (1). IGI Global, 2018; pp 15–34. DOI: 10.4018/IJFC.2018010102.

8. Botta, A.; de Donato, W.; Persico, V.; Pescapé, A. In *On the Integration of Cloud Computing and Internet of Things.* 2014 International Conference on Future Internet of Things and Cloud, 2014; pp 23–30. DOI: 10.1109/FiCloud.2014.14.

9. Cao, K.; Liu, Y.; Meng, G.; Sun, Q. An Overview on Edge Computing Research. *IEEE Access* **2020,** *8*, 85714–85728. DOI: 10.1109/ACCESS.2020.2991734.

10. Caprolu, M.; Di Pietro, R.; Lombardi, F.; Raponi, S. Edge Computing Perspectives: Architectures, Technologies, and Open Security Issues. In *2019 IEEE International*

Conference on Edge Computing (EDGE), 2019; pp 116–123. DOI: 10.1109/EDGE.2019.00035.

11. Wesley, D. "Microsoft Edge Identity Support and Configuration, 2021. https://docs.microsoft.com/en-us/deployedge/microsoft-edge-security-identity (accessed May 17, 2021)

12. Ying, G.; Liu, Y.; Jin, Y.; Chen, J.; Wu, H. A Novel Semi-Supervised Learning Approach for Network Intrusion Detection on Cloud-Based Robotic System. IEEE Access 2018, 6, 50927–50938. DOI: 10.1109/ACCESS.2018.2868171.

13. Gazzaz, S.; Nawab, F. In SoCC '19, Collaborative Edge-Cloud and Edge-Edge Video Analytics, Proceedings of the ACM Symposium on Cloud Computing; Association for Computing Machinery: New York, NY, USA, 2019. DOI: 10.1145/3357223.3366024.

14. Ting, H.; Ciftcioglu, E. N.; Wang, S.; Chan, K. S. Location Privacy in Mobile Edge Clouds: A Chaff-Based Approach. IEEE J. Select. Areas Commun. 2017, 35 (11), 2625–2636. DOI: 10.1109/JSAC.2017.2760179.

15. Huang, B.; Li, Y.; Li, Z.; Pan, L.; Wang, S.; Xu, Y.; Hu, H. Security and Cost-Aware Computation Offloading via Deep Reinforcement Learning in Mobile Edge Computing. Wirel. Commun. Mobile Comput. 2019, 2019, e3816237. DOI: 10.1155/2019/3816237.

16. Ibrahim, D. An Overview of Soft Computing. Procedia Comput. Sci. 12th International Conference on Application of Fuzzy Systems and Soft Computing, ICAFS 2016, Aug 29–30, 2016; Vienna, Austria, 2016, 102, 34–38. DOI: 10.1016/j.procs.2016.09.366.

17. Khan, I.; Belqasmi, F.; Glitho, R.; Crespi, N.; Morrow, M.; Polakos, P. Wireless Sensor Network Virtualization: A Survey. IEEE Commun. Surv. Tutor. 2016, 18 (1), 553–576. DOI: 10.1109/COMST.2015.2412971.

18. Guanwen, L.; Zhou, H.; Feng, B.; Li, G.; Li, T.; Xu, Q.; Quan, W. Fuzzy Theory Based Security Service Chaining for Sustainable Mobile-Edge Computing, Hindawi. Mobile Inf. Syst. 2017, 2017, e8098394. DOI: 10.1155/2017/8098394.

19. Hui, L.; Jing, T. A Lightweight Fine-Grained Searchable Encryption Scheme in Fog-Based Healthcare IoT Networks, Hindawi. Wirel. Commun. Mobile Comput. 2019, 2019, e1019767. DOI: 10.1155/2019/1019767.

20. Li, L.; Ota, K.; Dong, M. DeepNFV: A Lightweight Framework for Intelligent Edge Network Functions Virtualization. IEEE Netw. 2019, 33 (1), 136–41. DOI: 10.1109/MNET.2018.1700394.

21. Lima, M. V. M.; Ricardo, M. F. L.; Fernando, A. A. L. In A Multi-Perspective Methodology for Evaluating the Security Maturity of Data Centers, 2017 IEEE International Conference on Systems, Man, and Cybernetics (SMC), 2017; pp 1196–1201. DOI: 10.1109/SMC.2017.8122775.

22. Stephen, M. In Machine Learning: An Algorithmic Perspective, 2nd Ed.; Chapman and Hall/CRC: Boca Raton, 2014.

23. Merenda, M.; Porcaro, C.; Iero, D. Edge Machine Learning for AI-Enabled IoT Devices: A Review. Sensors 2020, 20 (9). DOI: 10.3390/s20092533.

24. Patel, M.; Sabella, D.; Sprecher, N.; Young, V. Mobile Edge Computing A Key Technology towards 5G; ETSI White Paper No. 11. In 16. ETSI White Paper, 11, First Ed.; 2015. https://www.etsi.org/images/files/etsiwhitepapers/etsi_wp11_mec_a_key_technology_towards_5g.pdf.

25. Rathore, S.; Sharma, P. K.; Sangaiah, A. K.; Park, J. J. A Hesitant Fuzzy Based Security Approach for Fog and Mobile-Edge Computing. *IEEE Access* **2018**, *6*, 688–701. DOI: 10.1109/ACCESS.2017.2774837.

26. Ray, P. P. A Survey of IoT Cloud Platforms. *Future Comput. Inform. J.* **2016**, *1* (1), 35–46. DOI: 10.1016/j.fcij.2017.02.001.

27. Rudas, I. J. Hybrid Systems. In *Encyclopedia of Information Systems*; Bidgoli, H.; Elsevier: New York, 2003; pp 563–70. DOI: 10.1016/B0-12-227240-4/00088-5.

28. Sethi, P.; Sarangi, S. R. Internet of Things: Architectures, Protocols, and Applications. *J. Electr. Comput. Eng.* **2017**, *2017*, e9324035. DOI: 10.1155/2017/9324035.

29. Sha, K.; Errabelly, R.; Wei, W.; Yang, T. A.; Wang, Z. In *EdgeSec: Design of an Edge Layer Security Service to Enhance IoT Security*, 2017 IEEE 1st International Conference on Fog and Edge Computing (ICFEC), 2017; pp 81–88. DOI: 10.1109/ICFEC.2017.7.

30. Subashini, S.; Kavitha, V. "A Survey on Security Issues in Service Delivery Models of Cloud Computing. *J. Netw. Comput. Appl.* **2011**, *34* (1), 1–11. DOI: 10.1016/j.jnca.2010.07.006.

31. Subramanian, E. K.; Tamilselvan, L. A Focus on Future Cloud: Machine Learning-Based Cloud Security. *Serv. Oriented Comput. Appl.* **2019**, *13* (3), 237–249. DOI: 10.1007/s11761-019-00270-0.

32. Sultana, T.; Khan, A. W. IoT-Guard: Event-Driven Fog-Based Video Surveillance System for Real-Time Security Management. *IEEE Access* **2019**, *7*, 134881–134894. doi:10.1109/ACCESS.2019.2941978.

33. Tanwar, S.; Bhatia, Q.; Patel, P.; Kumari, A.; Singh, P. K.; Hong, W. C. Machine Learning Adoption in Blockchain-Based Smart Applications: The Challenges, and a Way Forward. *IEEE Access* **2020**, *8*, 474–488. DOI: 10.1109/ACCESS.2019.2961372.

34. Tecuci, G. Artificial Intelligence. *WIREs Comput. Stat.* **2012**, *4* (2), 168–180. DOI: 10.1002/wics.200.

35. Vuong, T. P.; Loukas, G.; Gan, D.; Bezemskij, A. In *Decision Tree-Based Detection of Denial of Service and Command Injection Attacks on Robotic Vehicles*, 2015 IEEE International Workshop on Information Forensics and Security (WIFS), 2015; pp 1–6. DOI: 10.1109/WIFS.2015.7368559.

36. Steven, W. R.; Hammoudeh, M. Artificial Intelligence Agents as Mediators of Trustless Security Systems and Distributed Computing Applications; Computer Communications and Networks. In *Guide to Vulnerability Analysis for Computer Networks and Systems: An Artificial Intelligence Approach*; Parkinson, S.; Crampton, A.; Hill, R., Springer International Publishing: Cham, 2018; pp 131–55. DOI: 10.1007/978-3-319-92624-7_6.

37. Yang, W.; Wang, S.; Hu, J.; Zheng, G.; Yang, J.; Valli, C. Securing Deep Learning Based Edge Finger Vein Biometrics With Binary Decision Diagram. *IEEE Trans. Ind. Inform.* **2019**, *15* (7), 4244–4253. DOI: 10.1109/TII.2019.2900665.

38. Zhang, P.; Liu, J. K.; Yu, F. R.; Sookhak, M.; Au, M. H.; Luo, X. A Survey on Access Control in Fog Computing. *IEEE Commun. Mag.* **2018**, *56* (2), 144–149. DOI: 10.1109/MCOM.2018.1700333.

39. Zhao, X. P.; Jiang, R. Distributed Machine Learning Oriented Data Integrity Verification Scheme in Cloud Computing Environment. *IEEE Access* **2020**, *8*, 26372–26384. DOI: 10.1109/ACCESS.2020.2971519.

CHAPTER 6

SECURITY IN IOT USING ARTIFICIAL INTELLIGENCE

SANCHARI SAHA

Department of Computer Science and Engineering,
CMR Institute of Technology, Bengaluru, Karnataka, India

ABSTRACT

In various scenarios, data are referred to as the "oil of the twenty-first century," and the Internet-of-Things(IoT) as one of the most popular technologies. IoT is a sparse network in which sensors extract useful information from their surroundings and transmit it via the Internet. This IoT generation is vastly influencing our lives. Prior to the IoT, the majority of security threats were caused by information leakage or service disruption, however, with increased reliance on the IoT, security threats are directly mapped to our day-to-day lives and can easily affect even our physical security.

IoT is a brilliant technology for combating the COVID-19 pandemic and has the potential to meet significant challenges during a lockdown. This technology can detect real-time data as well as other pertinent information about COVID patients. However, the main concern in deploying the IoT in thiscurrent pandemic situation is data security and privacy, which is a matter of life and death from the perspective of patients' health.

As user privacy is of high importance in most IoT applications, a mechanism is required to support multi-platform security for different device

The Fusion of Artificial Intelligence and Soft Computing Techniques for Cybersecurity.
M. A. Jabbar, Sanju Tiwari, Subhendu Kumar Pani, & Stephen Huang (Eds.)
© 2024 Apple Academic Press, Inc. Co-published with CRC Press (Taylor & Francis)

interactions in IoT applications to prevent personal data from flowing to an unintended destination.

Artificial intelligence has paved the way for addressing various security issues that are present in IoT applications. In this chapter, readers will be able to visualize the role of artificial intelligence in correctly analyzing and preventing harmful attacks with greater efficiency.

6.1   INTRODUCTION

Internet-of-Things (IoT) has emerged as a cutting-edge technology and with its numerous applications it has gained popularity in both research and business domains. IoT can be thought of as a collection of various devices that require a cloud environment for data processing and transfer. Artificial intelligence (AI) is very much important in analyzing the stored data and providing reliable decisions in a short span of time. The assembled IoT devices communicate with one another via unique identifiers and share data via the Internet or cloud infrastructure.[1] Presently, the age where we are living, usage of AI and machine learning (ML) is critical in order to analyze cloud data quickly and accurately. Despite the fact that AI is playing a significant role in improving cyber security, cloud vulnerability, and IoT-based communications continue to pose a threat. Along with cloud and IoT security issues, attackers employ security threats related to AI also, which remains a threat to the cyberspace world. Furthermore, the vast majority of wireless IoT devices connected to public networks are vulnerable to ongoing cyber threats.

IoT communicates data in conjunction with big data and AI.[2] As per the article,[3] around 13 new devices are getting connected to the Internet every second, and the number of devices connected globally is expected to increase to more than 2 billion by 2025, nearly tripling from 2011. The greater the number of IoT devices, the greater the need for network security and cloud-based bigdata protection. Although the development and benefit of implementing an IoT network provide a high return on investment, one of the key security issues is the ongoing cyber threat to the security of IoT devices and cloud network infrastructure.

When considering the IoT as a collection of interconnected devices that provide intuitive Internet-based services such as data privacy and protection from various cyber-attacks during transportation, ensuring the highest

level of security is critical. AI or machine learning can be used to imple-
mentsupervised practice, unused practice, and reinforcement practice to
meet this requirement. The use of AI creates a better secure environment in
the cloud setup and ensures the possibilityof realizing the future of the IoT.

The entity-based architecture of IoT is split into three layers, namely
the terminal perception layer, network transport layer, and application
service layer as shown in Figure 6.1.[2] Terminal layer is responsible for
data absorption of IoT, whereas the network transport layer is responsible
to send information from the terminal to the application layer to perform
communication and connection-oriented functions. The application layer
on the other hand processes data from the network transport layer and
merges with different industries to support various IoT applications.[2]

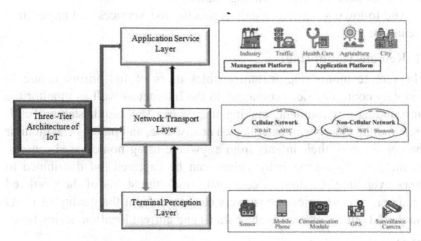

FIGURE 6.1   Entity-based architecture of IoT.
Source: Adapted from Ref. [2]

6.2 IOT SERVICES AND APPLICATION SCENARIOS

The primary goal of IoT is to create a global network of various types of
devices that can be connectedto the Internet for information exchange.
Interconnected devices can transfer data and perform various functions
without the need for human intervention when using IoT technology.
The IoT can be used ina variety of applications by installing sensors in
home appliances, medical equipment, and transportation, for example.

The sensor layer, network layer, and application layer comprise the IoT structure.[4]

Each of these layers serves a specific purpose. The sensor layer is in charge of sensing the environment and receiving data via wireless technology, radio-frequency identification, or sensor technology. All these technologies form the foundation of the IoT. The network layer is in charge of transferring data from the sensor layer to the application layer. In addition, the cloud framework is an important component of this layer for analyzing large amounts of data. The application layer is located at the top of the IoT architecture and provides specific services to customers based on processed and analyzed data. The IoT architecture can understand customer needs through these three layers and provide services that meet those needs while improving quality.

The following section describes specific IoT services and application scenarios:

(1) Remote Monitoring:

The remote monitoring feature enables users of IoT infrastructure to remotely control devices connected to the Internet as well as monitor the surrounding environment, making life easier. The health status of children and the elderly can be collected at any time, and parents can monitor the condition of their infants from anywhere using home-based sensors. Furthermore, real-time baby videos can be captured and distributed to users over time. Customers can easily track the status of their ordered goods or other logistics. Consumers can learn about the quality of items purchased online, as well as their status and current location, at any time.

(2) Home Automation:

Home automation[2] has been around for a long time, so it is not a surprising application concept for us. But what deserves our attention is that smart home appliances have become more versatile, intelligent, and capable of providing better service to users. When we enter our room from outside, the air conditioner automatically turns on and adjusts the temperature. There are numerous other home service products available that make human life more comfortable. Sweeping robots are another IoT-based home appliance that can turn on and off lights automatically without intervention. As a result, smart homes are a direct example of IoT services that aim to make human life easier and more comfortable.

(3) Prediction of Natural Disaster:

The IoT can help in predicting natural disasters such as floods, earthquakes, tsunamis, and droughts.[2] External sensors are used to collect data from the environment, process it, and reveal important information about potential natural disasters. This saves time in relocating people from disaster-prone areas while also minimizing damage to homes and other belongings as much as possible. The IoT has served as a solid infrastructure, propelling a revolution in many traditional offline industries. Even though the IoT is still in its early stages, it has gained popularity in a variety of application areas because it has positively impacted every aspect of humanity. Figure 6.2 highlights the usefulness of IoT in various domains.

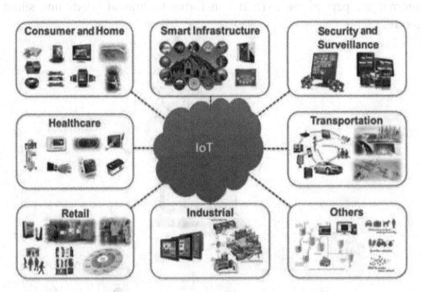

FIGURE 6.2   Usefulness of IoT in various domains.[2]

6.3   ASPECTS OF IOT IN COVID-19

The novel coronavirus epidemic that erupted in Wuhan, China in December 2019 has affected the entire world. The disease has been dubbed the Corona Virus Disease 2019 and has been named the world's most dangerous virus affecting human health. Currently, the global number of COVID-19 cases has surpassed 5,000,000, with approximately 350,000 people reported dead worldwide. This figure is rapidly increasing by an hour. Health workers and medical researchers have discovered a new methodto control the COVID-19 infection in this emergency medical situation. Continuous monitoring of this viral infection is critical for isolating infected patients and preventing relapse. In this situation, researchfields like AI, ML, and the IoT are churning out technologies that canbe used to address the increasingly important medical issues associated with COVID-19. Figure 6.3 is highlighting remote monitoring of patients using IoT.

The IoT has evolved into one of the most important technologies of the twenty-first century. In the modern era, we can connect cars, home appliances, health monitoring systems, and home security systems to the Internet with the help of IoT. Smart devices for exchanging data over the Internet are part of the IoT.[1] It transforms traditional goods into smart goods.

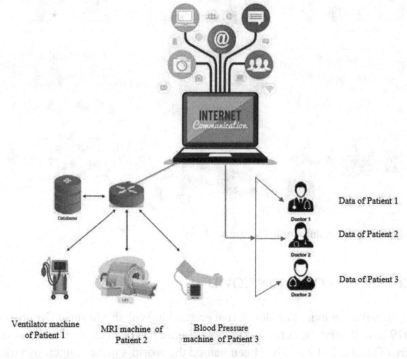

FIGURE 6.3   Remote monitoring of patients using IoT.

Source: Adapted from Ref. [29]

The goal of the IoT is to bring everything under the umbrella of this world's interconnected structure, allowing humans to not only control the devices around them, but also provide real-time updates on their states. IoT devices primarily perceive their surroundings and transmit data captured by them via the Internet without the need for human intervention. The IoT has emerged as an important part of human life, with a growing number of devices linked via IoT.[3]

The IoT has been developed as a promising technology that will play an important role in a variety of fields including healthcare, home automation, industrial automation, autonomous vehicles, smart grids, and so on. It ensures the best use of resources with low costs and modest negotiations in a seamless manner. As COVID-19 spreads around the world, it is critical to understand and clarify the role of the IoT in this pandemic. COVID-19 positive cases had surpassed 9 million by August 2020, with a 4.7% mortality rate.[4] Medical researchers are experimenting with and researching different solutions to help protect against COVID-19.[5]

The IoT is a collection of various components designed to serve multiple countries and combat the negative effects of COVID-19. With its scalable network, the IoT is capable of handling large amounts of data collected from sensor devices used by many applications to combat COVID-19. Furthermore, the IoT network's dependability reduces the time duration in delivering emergency information, which can aid in providing on-time response during this global pandemic. Because of the spread of this coronavirus, there has never been a greater need for the IoT.

The IoT has been developed as the most useful technology that can change our lives due to its powerful integration with seamless connection and other associated technologies. IoT applications in combating this global epidemic are widespread in many areas, and they play an important role in lowering the risk of coronavirus outbreaks.[11] The potential applications of IoT technologies that could be useful and necessary in combating COVID-19 are:

- Smart education
- Smart grid & AMI
- Block chain-based IoT
- AI-based accurate forecasting
- IoT enabled ambulances
- Automation
- Wearable devices

- Robot
- Mobile app
- Drone
- Smart gadgets
- IoT-based telemart
- Digital telehealth

6.4   SECURITY CHALLENGE OF IOT

Access control, device identification management, and data encryption are some of the approaches used in the IoT security procedure to reduce vulnerability. The IoT technology has grown in value over time, but while ensuring overall data perfection, the complexity of communication, identification, and product perception has increased. The increase in data size has an immediate impact on data and network vulnerabilities.[5]

The security of communication in an interconnected network and the connection of IoT devices pose a significant threat. The main concern with IoT devices in terms of data security is that most are incapable of dealing with cyber-attacks and privacy-related threats. As a result, it may endanger the entire IoT network. According to security experts, most IoT devices have inadequate security measures, making them an easy target for the attackers. In conclusion, ensuring security for large amounts of data, speed, diversity, and IoT infrastructure is a difficult task.[4]

The device identifier is a unique code that adds another layer of security. The identification codes for specific devices function similar to our mobile phone's IMEI number, but this does not apply to all connected objects in the IoT network. As a result, the current identification standard is applicable to all devices in the system.

When the exact physical location of the device is known, geographic location is another important factor that provides proper security. Although we can obtain such information from smart TVs or mobile phones, this is not true for all network-connected devices.

Access to other vulnerable devices also provides cybercriminals with easy access to other connected systems in the network. For example, a typical agreement with "Baby Monitor," which is linked to the IoT network, provides cybercriminals with easy access to other connected devices such as cars, homes, smart TVs, and many more. As a result, implementing the IoT as a safety device is a difficult task.

6.5 IMPLEMENTATION CHALLENGES OF IOT IN COVID-19 SCENARIO

Implementing IoT in the COVID-19 scenario imposes a number of challenges, which are highlighted below:

6.5.1 SCALABILITY

IoT devices are becoming increasingly popular as digital technology advances. The reason for this expansion is that IoT devices are not limited to a single App but are spread across multiple Apps that are popular today. According to a recent survey, the demand for smart home equipment will skyrocketbetween 2018 and 2022.

The major challenge in implementing IoT to combat the global pandemic of COVID-19 is scalability. Internet-of-Health-Technology (IoHT) alone necessitates a large number of devices in order to correctly interpret patients' vital signals and transmit them to the Internet cloud.

Currently, there are approximately 4.1 million active COVID cases worldwide. Each IoT instrument is made up of multiple sensors, making implementation of the IoT a difficult task. As infrastructure grows, a massive amount of data flows through these tiny IoT sensor nodes, increasing energy consumption.

6.5.2 BANDWIDTH LIMITATION

As the number of IoT devices are growing, the demand for bandwidth to send data from sensor nodes to the cloud also increasing. Presently, the majority of IoT devices make use of licensed spectrum provided by various mobile operators. However, along with this increasing number of devices, the demand for bandwidth, resulting data delivery delays and incorrect data transmission also are increasing.

When the number of devices in a fixed IoT infrastructure grows, Wi-Fi usage in the coverage area becomes unreliable. At present, most of the IoT devices utilize the 3G/4G/LTE networks in performing their functionalities.

The 3G/LTE/4G spectrum would soon be insufficient for a significant number of IoT users. During the COVID-19 pandemic, timely data transfer to entities connected to IoT devices was critical. Humanlives might be lost as a result of errors or data delays. Latency and low data rates can be overcome if the bandwidth is sufficient.

6.5.3   PRIVACY AND SECURITY

Because of the scalability and power limitations of the IoT devices, traditional cryptographic approaches are not feasible solutions to enforce protection. To ensure data protection, user privacy, and reliable authentication, security solutions must be energy efficient, and algorithms that securely identifyIoT networks must have minimal computational issues.[15] Therefore, lightweight protection algorithms must be developed in order to enforce security in IoT. The security requirements of IoT-enabled networks have increased as a result of the spread of coronavirus.

Following are the major concerns related to security while implementing IoT in COVID-19 scenario:

- The data sent from the sensors implanted on the COVID-19 patent's body must be correct.
- The data must reach to its intended location.
- There should be no duplication of data.
- Data stored in the memory of an IoT computer should not be accessible to the unauthenticated users.
- IoT devices with low computing capabilities should be considered for security.
- The requisite security algorithms must be accurate and maintain user confidentiality in addition to being lightweight.[9]

6.5.4   STRENGTH, DEFICIENCY, OPPORTUNITY, AND THREAT ANALYSIS

Table 6.1 shows IoT strength, deficiency, opportunity, and threat analysis review. Internal considerationsare restricted to companies or researchers who choose to incorporate IoT in every area. Internal variables can shift over time. External factors such as opportunities and risks are dependent on the market and cannot be adjusted.[16]

TABLE 6.1   Strength, Deficiency, Opportunity, and Threat Analysis.

Internal Factors	
Strength	**Deficiency**
Accuracy of data	High processing servers are required
On time treatment	Scalability of IoT device
Timely diagnosis	Huge data centers & data aggregation
Information of safety measure	Security & privacy preservations
High demand of IoT based systems	High bandwidth requirement
Accurate forecasting	Limited spectral resources
External Factors	
Opportunity	**Threat**
Creation of awareness about the requirement of IoT	Compatibility of devices
Creation of jobs	Use of unlicensed bands
Software defined radios	
Cooperative Communication	
Towards mm wave communication for higher bandwidths	

Consistency of the data in IoT plays a positive role for implementation in COVID-19' scenario. Sensors collect data from the atmosphere in real-time and send it to the cloud. It enables patients to receive prompt care potentially saving many lives. If anyone has COVID-19 symptoms and wants to see a doctor, the IoT can benefit by providing a telehealth network where medical advice can be obtained without having to go to a hospital or clinic. This applies to prompt COVID-19 diagnosis. IoT aids in the dissemination of information and security measures thereby helping prevention of the coronavirus.

There is a high demand for IoT-based systems due to the role of IoT in addressing the ongoing global coronavirus epidemic. Using AI in conjunction with IoT would enable researchers to better assess the need to combat COVID-19 in the future.

When IoT needs to be used to combat the virus, errors and bugs must be taken into account. Because of the large number of IoT devices and the scalability criteria, data processing units must have sufficient processing power. The data center should be large enough to store all of the patient records and documents.

The entire IoT network should be extremely stable, with security algorithms that are as simple as possible. Since most of the devices transmit data to the cloud on a regular basis, high bandwidth requirements are unavoidable. Mechanisms must be developed to make efficient use of limited spectrum.

It is not difficult to raise awareness about the use of IoT applications in eradication of coronavirus, particularly with the advancement of digital technology and the widespread use of smartphones. Furthermore, IoT assists the industry in creating jobs in local markets and effectively contributes to the growth of every country's economy.

6.6   SOLUTION TO FIGHT AGAINST COVID-19

Problems in implementing scalable IoT network in COVID-19 context are controversial, but there are solutions for these challenges in the literature that can assist in the effective operation of IoT networks.

Because of their scalability, most IoT devices are compact and easy to use. Steps must be taken to ensure that data are accessible at the destination in a secure and efficient manner. Since IoT nodes are mostly not physically secured, data protection serves as the backbone in operating IoT networks. If proper authentication protocols are not used, data can be easily duplicated. Primitive basic attack recognition, channel state masking, intrusion detection, localization, and data verification are all included in protection. A single data change will result in major issues.

During COVID-19's global epidemic, errors in a patient's medical health reports and the smart grid power produced by IoT devices sent to physicians can cause major problems. Traditional cryptographic approaches are not feasible solutions due to the power limitations of IoT devices. End-to-end content safety, user authentication, and customer trust in the IoT environment are all dependent on energy-efficient encryption that uses less memory and has a lower computational complexity.

Easy encryption methods are at the heart of lightweight security algorithms. Various metrics, such as arrival time, step-by-step details, and received signal strength indicators, can be used to build lightweight security algorithms for IoT. The contact of IoT devices gave rise to linked fingerprints. These linked fingerprints are encoded with a symmetric key and sent to a server, where Pearson correlation coefficient is computed using the linked fingerprint of the linked IoT system.

The Pearson correlation coefficient is a straightforward method for detecting negativity in IoT network.[9] Table 6.2 summarizes some of the most significant lightweight security algorithms in the literature.[18–21]

TABLE 6.2   Existing Security Algorithms for IoT.

Security Requirements	Gope et al. [18]	Dong et al. [19]	Ali et al. [20]	Kamal et al. [21]
MITM	YES	YES	YES	YES
	YES	NO	YES	YES
	NO	NO	YES	YES
	YES	NO	NO	NO
	NO	NO	NO	YES
	NO	NO	YES	YES

6.7   ENCRYPTION METHODS IN PRACTICE

6.7.1   ANONYMIZATION PROCESS

Anonymous technology is used to mask the identities of participants in a group of people by eliminating explicit attributes such as usernames and ID numbers. The most important data we need during transmission are not wasted because of the lost information that is linked to the user's specific identity. Several privacy-oriented models such as K-anonymous, L-variant,[22] and T-proximity[23] models are used in various scenarios. Taxonomic identifiers, semi-identifier attribute sets, and responsive properties are the three types of participant characteristics in the K- anonymous model.

The explicit identifiers are extracted from the data before it is posted, and the data in the quasi-identifier attribute set are normalized to ensure that at least k records have a single partial-identifier. In the k- anonymous strategy, by linking or comparing data to other context data or looking at unique features contained in the published data, attackers can re-identify victims.[24] In order to defend against this flaw, some researchers suggested I-variation and t-proximal models based on the k-anonymity technique.

The diversity of sensitive features in each semi-identifier class should not be less than the I-variation value, lowering the likelihood of sensitive features and their owners getting matched. In each equivalent class to the T-Closeness module, the distance between the distribution of sensitive properties and the normal distribution of sensitive properties should not exceed the upper limit T.

6.7.2   CRYPTOGRAPHIC TECHNIQUE

Until uploading data to a cloud server, cryptographic methods can encrypt the data. Cryptographic approaches necessitate a large computational overhead (millions of times that of multiplier projection) as well as effective and efficient key management.[17]

When it comes to generating important pairs based on certain mathematical problems, homomorphic encryption (HE) technology is more secure. Computers are the most difficult to crack. Generally, the public key and the private key build the key pair. Third parties are given access to the public key and certain process actions. The third party can conduct all operations on the encrypted data and return the result, which is only represented by the private key. Hence, all information in the process is kept private. The Rivest–Shamir–Adleman (RSA) and Elliptic Curve Cryptography (ECC) algorithms are two common homomorphic encryption algorithms.

6.7.3   INTERRUPTION OF DATA

Example methods for global module training that obscure data models are, pair side disturbances and qualitative disturbances. Differential privacy (DP), which is used to gather information without disclosing a particular entry, is often linked to additive disruption. The adversary cannot differentiate between the outputs of the neighboring dataset due to DP's mechanism that protects the neighboring dataset's various records. Certain mechanisms, such as Laplacian, Exponential, and Intermediate mechanisms, are used by DP to add noise to data.

The Laplacian mechanism identifies DP protection by applying random noise to the exact query product using the Laplacian distribution. Unlike the Laplacian process, the Exponential scheme chooses the best product for each question based on the likelihood. Random coefficient is a type of data manipulation technique that is used to manipulate data qualitatively.

The random projection scheme tries to constructa new data representation by using random measurements to minimize the count of measurements. In general, data opacity is used in data mining to protect the privacy of users when collecting high-quality data.

6.8 USE OF ARTIFICIAL INTELLIGENCE IN IOT

As security threats in the modern environment continue to evolve, it is critical to incorporate AI into security strategy of design and maintenance. Because of the sophistication of modern cyber threats including scarcity of cyber security expertise, network security teams rely on machine learning and other AI-based technologies to track, defend, and mitigate advanced attacks.[25]

While businesses use AI to improve their security, cybercriminals use active software creation, automation, and ML to exploit network vulnerabilities, detect, and exploit AI.[26]

Cybercriminals now have the opportunity and ability to conduct rapid and complex attacks that enable malicious devices to penetrate corporate networks due to the number and diversity of IoT devices that infiltrate network infrastructure. The risks of today's digital transformation efforts are exacerbated by AI's potential attack capability.

As a result, AI will soon have a way to successfully defend or threaten IoT, potentially putting cybersecurity experts and cybercriminals in an arms race.

IT teams must consider recent developments in cyber strategy that will lead to an AI-based threat landscape in the coming years in order to defend digital changes and maintain a strict security posture. They must also know AI capabilities to allow inside their security stock in order to maintain a consistent security posture as and when their network develops and expands.

6.8.1 CHANGED AI-DRIVEN COERCION LANDSCAPE

Cybercriminals have also started to use automated and scripted methods to improve the pace and sophistication of their attacks. Because of these advanced capabilities, the increase in revenue from quarter 1 to quarter 2, 2018 was 240%. Cybercriminals are setting the groundwork for AI to automatically map networks, evaluate threats, choose attack vectors, deploy them and also carrying out tests in order to create personalized and automated attacks.

Cybercriminals are going through their own digital revolution and as a result, are still using manual threat detection and active development to speed up malware development and defeat advanced security solutions. The final adoption of AI will fasten this process even further.

Because of the rapid advancement of cyber malware and exploitative developers, the raising question is when the business will be targeted. Unfortunately, many businesses continue to rely on the Legacy Point Product Solutions, which on average have more than 30 separate items in their networks to effectively recognize and react to today's sophisticated attack strategies. Simply put, as cybersecurity skills deteriorate, manual threat detection, and reliance on security tactics as well as today's cybercriminals' advanced capabilities, cannot be reconciled.

6.9   ANALYZING FEASIBILITY OF ARTIFICIAL INTELLIGENCE IN MAKING IOT MORE SECURE

According to the overview of the various security issues faced by IoT infrastructure, it is found that most of the security issues of IoT share same features that make security in IoT more complicated and create unique criteria for security protection in other areas.[2]

In the usual IoT sense, data delivery is relatively stable. So, detecting suspicious activity and data outliers becomes a significant IoT security requirement.[2] In general, IoT systems can only perform basic taskslike data collection and transmission due to a lack of resources.

A large number of common devices operate in a continuous business mode and distribute collected datain a reasonably consistent manner. Static modes are common in network traffic on user IoT devices. Static signals are sent by these devices to a small number of endpoints, allowing for more robust, and organized network operations. Whereas denial of service (DoS)/distributed denial of service (DDoS) attack on the other hand, produce slightly diverse network traffic compared to IoT devices. As a result, real-time data tracking and classification, as well as time recording, and prior warning of suspicious commercial data sources are successful security measures against a variety of security threats.

There is a lack of prior expertise due to the unpredictability and diversity of attack modes and it meetsthe need for the security model's rhythm and generalization capacity.

Personal privacy had previously placed information security in jeopardy. With the growth of the IoT industry, attack scenarios for the IoT are changing as well. There are countless attacks on hardware and software bugs, communication interfaces, and cloud services. Defenders, on the other hand, have no previous knowledge of new attack methods and are unable to respond quickly.

Only after they have suffered a loss will comprehend the attack, and thereby significantly increasing the users' security risk. Without prior experience, the security model must be able to sustain impact in a number of unknown situations that necessitate high security, generalization, and data control.

The security approach of low-power equipment and microsystems terminals limits the ability to automate security. Complex security procedures are not feasible because large IoT system integration necessitates optimized power and low-cost solutions.

As a result, new IoT security practices have a hard time keeping up with older security techniques. Many of them lack the capacity to regenerate and grow on their own, which can be resolved by "activeimmunity" or "auto-immunity." These methods depend on humans to maintain and upgrade databases, identify new attack modes, and intercept rules in the absence of initiative. The majority of manual participation lags behind in terms of security and is unable to learn the security plan in a timely manner. The question of how to build automated protection mechanisms with "active immunity" has become a pressing issue in IoT security. IoT security plans will efficiently handle vast amounts of data and can be implemented on a large-scale and on robust basis.

IoT is a multi-layer framework that protects overall security concerns while also making data more dynamic and large-scale. It runs on terminals, networks, and service platforms. The unity and credibility of IoT have created an immediate need for security solutions that can process vast amount of data, as well as the need to assure that solutions may be applied fairly on large scale basis, and that they can be successful and developed in complex situations according to new situations, at any moment.

6.10 ARTIFICIAL INTELLIGENCE BASED SECURITY TOOL

The massive existence of IoT devices has given the computing world a new dimension and design transition. The number of connected devices at each home is increasing at alarming rate, necessitating more secure cyberspace infrastructure to mitigate the risk of leisure data, data in use, and data in motion as the most basic security requirements. Furthermore, IoT data collection, communication routes, and the cloud network on which data storage and analysis takes place, all follow stringent security criteria.

The implementation of various AI frameworks offers more complex and functional interiors due to the existence of its highly flexible cyber-physical structure and the interconnected devices whose data movements and analysis take place through very complex large area networks. It combines the features of Intrusion Detection System (IDS) and Intrusion Prevention System (IPS). Many businesses have implemented AI as key part of their security intelligence system, with the aim of ensuring that their infrastructure has a stable cyber defense posture to minimize risk.

Due to a lack of conventional security technologies that are heavily focused on rules, AI has emerged as a key component in maintaining cyber security. In the continuing barrage of cyber threats, AI is currently supporting agencies for massive security control assistance. Although AI is not bulletproof, its implementation is becoming the default standard as part of cyber defense strategy and is providing excellent support on a much higher level. All IoT data storage and computing exist in cloud environments, and cloud protection is another important factor. Although AI is not bulletproof, its application is becoming the default standard as part of cyber defense strategy and is providing excellentsupport on a much higher level.

The key benefits of AI/ML delivery include, but are not limited to real-time monitoring of existing vulnerabilities, big IoT data analytics, cyber-attack detection, and content delivery threat alerts.[27]

While AI/ML advancements promise to be a big boon to robotics science and cybercity, they also have the opposite effect, allowing hackers to fully develop AI/ML solutions for cyber purposes.[28] They are working 24/7 a week to develop and distribute more advanced AI/ML solutions that will use new attackvectors and promote similar attacks.

In addition, hackers are using AI/MI to test and improve their own malware, as well as merge and dilutetheir opponent's infrastructure with other AI/ML solutions.[25] With an increased count of AI/ML testing, the negative impact also gets increased. Other drawbacks of AI include its high cost, realism range, inability to represent human beings, unemployment, the need for human feedback to develop, and the effectiveness of its response to various cyber-attacks. The accuracy and availability of its training dataset, which comes from a variety of sources, determines its effectiveness.

Since it lacks ingenuity, expertise, and development, it needs precise datasets to learn from the necessary level. Because of malicious internal or external threats, technology is becoming more widely used; training data can be accurately exploited to build algorithms with disastrous consequences and harmful flaws. Several studies[9,10] have suggested various solutions for detecting and preventing malware, including non-ML solutions, in the field of IoT security.

As a result, a hybrid taxonomy and DDoS attack protector for SDN-based cloud environments using the history-based IP filtering (eHIPF) scheme are developed. In conclusion, this study focuses onnon-ML-based solutions, and research in this area appears to be successful in evaluating DDoS detection and prevention in SDO environments. The approach presented here addresses IoT security issues to deal with the cyber threat that is best solved in the cloud computing setting.

The training procedure is determined by the raw binary stream characteristics of the models.[27] Experiments have shown that the proposed approach is 75% efficient. However, there is still further work to be done in order to find their solution so that their research into evaluating their skill is limitedto the problem.

According to the authors of stream-based malware recognition using sensory neural network research,[13] automatic malware detection using automated neural networks (CNN), and other ML algorithms may be useful.[28] They used CNN, multi-layer perceptron (MLP), support vector machine (SVM), and random forest (RF) for classification, with an accuracy of 85% forall groups in their research using CNN and RF, demonstrating accuracy and recall. We may deduce from this that more methodology is needed to obtain a more accurate result. We may discover from the related articles that the malware detection machine on the host side uses non-algorithms with a maximum accuracy of 85%.

The study examines the role of AI and ML in combating emerging cyber threats (including malware) in cloud computing environments, especially in the context of IoT security. It uses the various solution approaches to fight and reduce IoT cyber threats on host- based and network-level cloud computing environments.

6.11 INSIGHTS OF AI-BASED SECURITY SOLUTION

A computerized device or its configuration does not provide security by itself.[14] Protection still extends to computer technology,[15] as cyber-criminals and hackers continue to develop new methods and techniques to discover vulnerabilities throughout our networks. To provide defense against these threats, a more complex and sensitive framework is needed. These solutions, according to security experts, should provide AI and ML with impenetrable security policies as they collect and analyze data from previous attacks and provide solutions based on that data.[25] These devices continuously monitor the network and look for past and potential attacks.

As a result, AI/ML solutions don't wait for an attack to happen; instead, they work to anticipate attacks based on the past and provide solutions to combat the threat. While hackers invent new technologies every day, no attack is completely unique; rather, a slightly modified version of old methods is used. Most of the security solutions with IoT devices are based on the fact that our AI/ML solutions are equipped with attack history and prototypes and can easily identify future attack wall defenses.

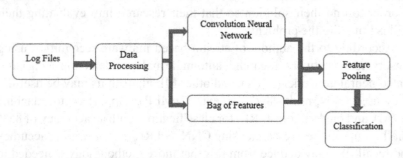

FIGURE 6.4   Data classification architecture.

Furthermore, since AI and ML operate without human interference, there is no need for physical resources to track the network 24/7 days a week. Hiring a large number of cyber security experts saves businesses a lot of money. When there is a large database to deal with, ML becomes very much involved.[26] The study uses an obligatory dataset to execute user data, complete the log file for network level recognition of the host base and network data in order to put the algorithm into operation. Figure 6.4 shows the data classification architecture. Although the number of security warnings and alerts can be overwhelming for humans, sophisticated security systems applications that use AI/ML can help.

It would be almost impossible for security teams in large cloud data centers to maintain security without the assistance of these sophisticated security systems. By developing predictive analytics to avoid future research, this study addresses security concerns on cloud network infrastructure.

IoT security is a crucial part of implementing reactive and preventative security policies that apply control to physical networks and software layers. To help distinguish or classify attacks, the following structure employs extensive practice.

System authentication, intrusion detection, DoS/DDoS attack detection and malware detection are all taxonomy tasks.[2] AI solutions must be able to correctly distinguish approved and unauthorized devices for system authentication; solutions must also be able to categorize common and unusual network behaviors for intrusion detection; and so on.[27,28] The current solutions for these issues are reviewed and summarized the flow of most of them in the process depicted in the Figure 6.5.

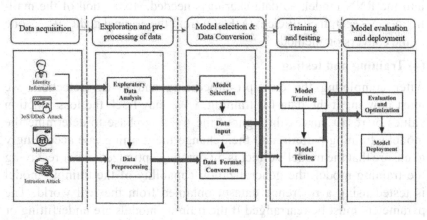

FIGURE 6.5   AI solution steps for IoT security.

Source: Adapted from Ref. [2]

(1) Data acquisition

Machine Learning solutions typically require datasets from specific environments.[2] To evaluate training datasets, you must select the appropriate data collection setting for various issues. To account for user variations, the dataset for system authentication should include information on device configuration, user behavior, and operating habits.

(2) Exploration and pre-processing of data

The effectiveness of an approach is directly related to the consistency of the training data.[2] The data for the IoT come from a variety of sensors in various locations. The initial data collection, on the other hand, had a number of flaws, including a skewed data distribution and missing data.

(3) Model selection and data Conversion

Many ML models are used to secure IoT, but each model has its own scenario, so the best model should be chosen based on the model's features and issues. The size of the dataset and the data's pre-search performance are primary factors in model selection.[2] For example, if the dataset contains more complex models and training must be completed quickly, a lightweight algorithm such as naive Bayes can produce the expected results.[2]

Data collected in real-world applications differ from the input data needed by the model, and it must be adjusted to suit the model's requirements.

Audio data obtained by voice sensors, cannot be directly entered into the RNN model, so data sharing is needed.[2] Extraction of the mail-frequency spectral coefficient (MFCC) from the original audio data is one of the conversion methods.

(4) Training and testing

After completing the data pre-processing and model selection, we must now input the data for training. [2] We may track the loss function values or result curves throughout the training phase to determine the model's training pattern, and then change the learning rate accordingly to ensure that the model's effect is moderately optimized. After receiving the training model, the generalization capability of the training model is tested using a reference dataset obtained from the real world. The parameters must be rearranged if the training models are underfitting or overfitting .

(5) Model evaluation and deployment

We may use some impact measures to compare various samples after training when choosing the final design for actual deployment and implementation.

6.12 FUTURE RESEARCH DIRECTION

AI and ML have a major impact on the growth of cyberspace. Similarly, IoT computers, smart houses, smart cars, and other innovations increase our everyday quality of life. An advanced protection approach isn't complete unless it includes AI and ML components. AI and ML solutions mainly help distinguish correlations between previous attacks and provide instant warning when another of the same pattern is detected. The best thing about AI and ML is that it can continuously understand user's behavior, evolving habits, and various manipulations.

According to one of the research recommendations agreed by security experts, standardized available datasets make it easier for ML-based solutions to interpret and analyze data. ML-based systems can be very useful in combating cyber threats once the dataset is specified and standardized.

It can be recommended to use an unsupervised solution or drawing a fine line between the controlled ones based on the features derived from data collection and based on the study solution. AI and ML systems will function without human involvement or making limited human intervention producing more balanced and efficient systems.

Although the scope of the research is limited to host-based and network-level cloud computing environments, IoT is a hybrid paradigm for combating and mitigating cyber threats, and a variety of algorithms can be employed to achieve this.

6.13 CHAPTER SUMMARY

We live in a time when AI and ML are integral parts of some of the most innovative security solutions. AI and ML play a significant role in not only enhancing conventional cyber security, but also in improving our everyday quality of life through IoT devices such as smart homes, smart cars, and other connected devices. AI and ML systems will function without human involvement, making limited human intervention systems more balanced and efficient. One of the most basic recommendations made by security professionals is to standardize the available datasets in order to facilitate data-based solutions and allow data to be quickly understood and analyzed. Once the dataset has been defined and standardized, ML-based systems can be extremely useful in combating all types of cyber threats.

ACKNOWLEDGMENT

With a profound sense of gratitude and respect, I am thankful to Dr. Dinesh K Anvekar, professor, Dept.of ECE, RNS Institute of Technology, Bengaluru for his contribution of time in reviewing the content and sharing valuable comments & suggestions that has helped me to widen my knowledge in writing this book chapter.

KEYWORDS

- IoT
- security
- artificial intelligence
- COVID-19
- challenges
- privacy

REFERENCES

1. Girma, A. Analysis of Security Vulnerability and Analytics of IoT (IOT) Platform. *Inf. Technol. New Generat.* **2018**, *738*, 101–104.
2. Wu, H.; Han, H.; Wang, X.; Sun, S. Research on Artificial Intelligence Enhancing Internet of Things Security: A Survey. *IEEE Access* **2020**.
3. Patel, M. McKinsey & Company, Jan 13, 2020 [Online]. https://www.mckinsey.com/industries/semiconductors/ourinsights/ whats-new-with-the-internet-of- things (accessed Apr 20, 2020).
4. Growth Enabler IoT, Market Pulse Report, IoT (IoT), April 2017 [Online]. https://growthenabler.com/flipbook/pdf/IOT%20Report.pdf.

5. Yadav, E. P.; Mittal, E. A.; Yadav, H. IoT: In *Challenges and Issues in Indian Perspective*, 2018 3rd International Conference on IoT: Smart Innovation and Usages (IoT-SIU); Bhimtal, India, 2018.

6. Williamson, J. Dummies, 2020 [Online]. https://www.dummies.com/careers/find-a-job/the-4-vs-ofbig- data/ (accessed Apr 26, 2020).

7. Efstathopoulos, P. 2019, July 29, 2019 [Online]. https://www.nortonlifelock.com/ blogs/rese arch-group/cloudsecurity- overwhelming-ai-and-machine-learning-can-help (accessed April 26, 2020).

8. Dataflair Team, Sept 15, 2018 [Online]. https://data-flair.training/blogs/how-iot-works/ (accessed Apr 20, 2020).

9. Yadav, E. P.; Mittal, E. A.; Yadav, D. H. In *IoT: Challenges and Issues in Indian Perspective*, 3rd International Conference on IoT: Smart Innovation and Usages (IoT-SIU); Bhimtal, India, 2018.

10. Wu, L.; Ping, R.; Ke, L.; Hai-Xin, D. In *Behavior-Based Malware Analysis and Detection*, 2011 First International Workshop on Complexity and Data Mining, 2011.

11. Saeed, N.; Bader, A.; l-Naffouri, T. Y.; Alouini, Md. S. When Wireless Communication Faces Covid-19: Combating the Pandemic and Saving the Economy. arXiv preprint arXiv:2005.06637, 2020.

12. Anderson, H. S.; Kharkar, A.; Filar, B. Evading Machine Learning Malware Detection. *Int. J. Appl. Eng. Res.* **2017**.

13. Yeo, M.; Koo, Y.; Yoon, Y.; Hwang, T.; Ryu, J.; Song, J. In *Flow-Based Malware Detection Using Convolutional Neural Network*, 2018 International Conference on Information Networking (ICOIN); Chiang Mai, 2018.

14. Chung, B.; Kim, J.; Jeon, Y. In *On-demand Security Configuration for IoT Devices*, 2016 International Conference on Information and Communication Technology Convergence (ICTC); IEEE, 2016.

15. Cano, J. J. *ISACA J.* Sept 01, 2016 [Online]. https://www.isaca.org/resources/isaca-journal/issues/2016/volume-5/cyberattacksthe-instability-of-security-and-control-knowledge (accessed Apr 04, 2020).

16. Berry, T. How to do a Swot Analysis for Better Strategic Planning [Online]. https:// articles.bplans.com/ how-to-perform-swot-analysis/.

17. Ghanchi, J. March 19, 2019 [Online]. https://thenewstack.io/the-possibilities-of-ai-and- machine-learning-for-cybersecurity/ (accessed Mar 30, 2020).

18. Gope, P.; Sikdar, B. Lightweight and Privacy-Preserving Two-Factor Authentication Scheme for Iot Devices. *IEEE IoT J.* **2019**, *6* (1), 580–589.

19. Dong, Z. C.; Espejo, R.; Wan, Y.; Zhuang, W. Detecting and Locating Man- in-the-Middle Attacks in Fixed Wireless Networks. *J. Comput. Inf. Technol.* **2015**, *23* (4), 283–293.

20. Ali, S. T.; Sivaraman, V.; Ostry, D.; Tsudik, G.; Jha, S. Securing First-Hop data Provenance for Bodyworn Devices using Wireless Link Fingerprints. *IEEE Trans. Inf. Forensics Secur.* **2014**, *9* (12), 2193–2204.

21. Kamal, M.; et al. Light-Weight Security and Data Provenance for Multi-hop IoT. *IEEE Access* **2018**, *6*, 34439–34448.

22. Zhou, B.; Pei, J. e k-anonymity and l-diversity Approaches for Privacy Preservation In Social Networks Against Neighborhood Attacks. *Knowl. Inf. Syst.* **2011**, (1), 47–77.

23. Li, N.; Li, T.; Venkatasubramanian, S. In *t-closeness: Privacy Be-yond k-anonymity and l-diversity*, Proceedings of the 12 Security and Communication Networks IEEE 23rd International Conference on Data Engineering; IEEE, Istanbul, Turkey, April 2007; pp 106–115.

24. Sweeney, L. k-anonymity: A Model for Protecting Privacy. *Int. J. Uncertain. Fuzziness Knowledge Based Syst.* **2002,** *10* (5), 557–570.

25. Mishra, S.; Sagban, R.; Yakoob, A.; Gandhi, N. Swarm Intelligence in Anomaly Detection Systems: An Overview. *Int. J. Comput. Appl.* **2021,** *43* (2), 109–118.

26. Gaurav, D.; Tiwari, S. M.; Goyal, A.; Gandhi, N.; Abraham, A. Machine Intelligence-Based Algorithms for Spam Filtering on Document Labeling. *Soft Comput.* **2020,** *24* (13), 9625–9638.

27. Dwivedi, R.; Dey, S.; Chakraborty, C.; Tiwari, S. Grape Disease Detection Network based on Multi-task Learning and Attention Features. *IEEE Sens. J.* **2021.**

28. Gaurav, D.; Shandilya, S.; Tiwari, S.; Goyal, A. In *A Machine Learning Method for Recognizing Invasive Content in Memes*, Iberoamerican Knowledge Graphs and Semantic Web Conference; Springer: Cham, Nov 2020; pp 195–213.

29. Kamal, Mohsin, Abdulah Aljohani, and Eisa Alanazi. "IoT meets COVID-19: status, challenges, and opportunities." arXiv preprint arXiv:2007.12268 (2020).

CHAPTER 7

CYBER SECURITY FOR INTELLIGENT SYSTEMS

ABINAYA INBAMANI,[1] N. DIVYA,[1] R. R. RUBIA GANDHI,[1]
M. KARTHIK,[1] and M. SIVARAM KUMAR[2]

[1]*Sri Ramakrishna Engineering College, Coimbatore,
Tamil Nadu, India*

[2]*Karpagam Academy of Higher Education, Coimbatore,
Tamil Nadu, India*

ABSTRACT

The utilization of the Internet of things (IoT) has expanded dramatically, with online protection concerns expanded alongside it. On the bleeding edge of online protection is artificial intelligence (AI), which is utilized for the improvement of intricate calculations to ensure organizations and frameworks, including IoT frameworks. Notwithstanding, digital aggressors have figured out how to take advantage of AI and have even started to utilize antagonistic AI to complete network safety assaults. This audit chapter aggregates data from a few other overviews and exploration papers in regards to IoT, AI, and assaults with what is more, against AI and investigates the connection between these three subjects with the motivation behind extensively introducing and summing up significant writing in these fields.

The Fusion of Artificial Intelligence and Soft Computing Techniques for Cybersecurity.
M. A. Jabbar, Sanju Tiwari, Subhendu Kumar Pani, & Stephen Huang (Eds.)
© 2024 Apple Academic Press, Inc. Co-published with CRC Press (Taylor & Francis)

7.1   INTRODUCTION

Internet of Things (IoT) was conceived[1] in 2008, its development has been blasting, and presently, IoT is a section of day-to-day existence and has a spot in many homes and organizations. IoT is difficult to define as it has been advancing and evolving since its origination; however, it very well may be best perceived as an organization of advanced and simple machines and processing gadgets furnished with novel identifiers (UIDs) that can trade information without human mediation.[2] A human interfacing with a focal center point gadget or application, frequently a portable application, that then, at that point goes on to send information and directions to one or numerous periphery IoT gadgets.[3] The periphery gadgets can finish capacities whenever required and send information back to the center point gadget or application, which the human would then be able to see. The IoT idea has given the world a more elevated level of openness, uprightness, accessibility, adaptability, confidentiality, and interoperability as far as gadget availability.[4] Nonetheless, IoTs are helpless against cyber attacks because of a mix of their different assault surfaces and their originality and in this manner, the absence of safety normalizations and prerequisites.[5] There is an enormous assortment of cyber attacks that aggressors can use against IoTs, contingent upon what part of the framework they are focusing on and what they desire to acquire from the assault. All things considered, there is an enormous volume of investigation into network safety encompassing IoT. This incorporates artificial intelligence (AI) ways to deal with shielding IoT frameworks from assailants, ordinarily in the wording of distinguishing strange conduct that might demonstrate an assault is happening.[6] Nonetheless, on account of IoT, digital aggressors continuously have the advantage as they just need to find one weakness while network safety specialists should secure various targets. This has prompted expanded utilization of AI by digital aggressors too, to impede the muddled calculations that distinguish odd action and pass by unseen.[7] AI has gotten a lot of consideration with the development of IoT advancements. With this development, AI innovations, for example, choice trees, straight relapse, AI, support vector machines, also, neural organizations, have been utilized in IoT network safety applications to be ready to distinguish dangers and possible assaults. Researchers in Ref.[8] give a complete audit of the security hazards identified with IoT applications and potential balances just as analyze IoT advancements as far as uprightness, obscurity, confidentiality, protection, access control, confirmation, approval, strength, and self-association. The creators propose profound learning models utilizing CICIDS2017 datasets for DDoS assault recognition for online protection in IoT (Internet of things), which give high precision, that is, 97.16%.[9] In Ref. [10], the creators assess the artificial neural networks (ANN) in a door gadget to be ready to recognize inconsistencies in the information sent from the edge gadgets. The outcomes show that the proposed approach can work on the security of IoT frameworks. The creators in Ref. [11] propose an AI-based control approach for location and assessment just as a pay of digital assaults in modern IoT frameworks. In Ref. [12], the creators give a vigorous unavoidable discovery to IoT environments, furthermore, foster an assortment of ill-disposed assaults and guard components against them just as they approve their methodology. In Ref. [13], the creators investigate the new advancement of AI decision-making in digital actual frameworks and find that such development is basically independent because of the expanding incorporation of IoT gadgets in digital actual frameworks, and the worth of AI dynamic because of its speed and efficiency in dealing with huge heaps of information is logical going to make this advancement inescapable. The creators of Ref. [14] examine new ways to deal with hazard investigation utilizing AI and AI, especially in IoT networks present in industry settings. At long last, Ref. [15] talks about techniques for catching and surveying network protection dangers to IoT gadgets to normalize such practices so that danger in IoT frameworks might be all the more efficiently identified and ensured against. This survey paper covers an assortment of points in regards to network protection, the Internet of things (IoT), AI (computer based intelligence), and how they all identify with one another in three overview style segments and gives a complete audit of cyber attacks against IoT gadgets just as it also suggests AI-based strategies for securing against these assaults. The extreme objective of this paper is to make an asset for other people who are exploring these pervasive points by introducing synopses making associations.

7.2 METHODS OF ATTACKS IN IOT DEVICES

When the devices are connected to internet, it is prone to lot of cyber attacks. The attacks can be either in the hardware or the software or the interface. Identification of the surface is important before rectifying it. Figure 7.1 shows the basic level of interfacing the wireless sensor node with the web interface. The data collection and transmission layer, gateway, database, and web interface along with the monitoring and visualization center should be protected from cyber attacks. The wireless sensor nodes are called motes in which the inbuilt sensors of the motes transmit the information to the database through the processor and the gateway. The gateway has a remote monitoring station for visualization and monitoring of the data.

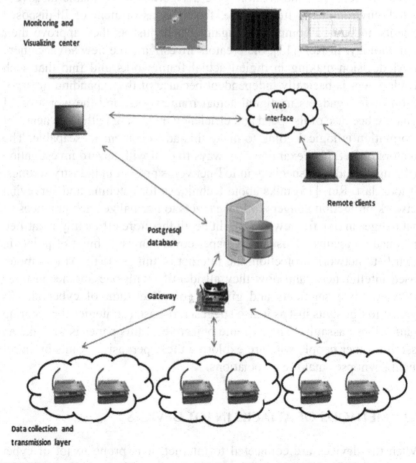

FIGURE 7.1　Basic wireless sensor network integration with internet of things.

Figure 7.2 shows the IoT devices connected to the cloud data server via network gateway. The user can visualize the data using a cloud data server.

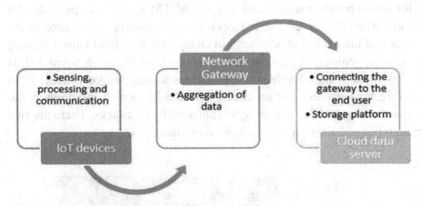

FIGURE 7.2 Process of communication.

Physical attacks are the lowest type category attacks. The hardware of the device is attacked. Outage attacks are the best example. The entire functionality of the network is affected by injecting false codes, injecting virus or malware into the device with the help of USB. Signal jammers are also considered as one of the best examples of physical attacks.

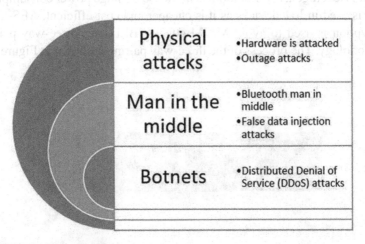

FIGURE 7.3 Types of attacks.

The signals blocked will lead to misinterpretation of results leading to false signal injection. MITM is another famous method of attack in networks. The Man in the Middle Attack acts like a proxy and changes the information between sender and receiver. MITM attacks take place amidst the client and the server. The IoT devices send sensitive information to the server and the MITM attacks tend to change the threshold values leading to mis-happenings. Hence, utmost care needs to be taken when IoT is incorporated in the HealthCare applications as they are very sensitive to the threshold values. The absence of standardization of protocols is the major drawbacks of IoT leading to vulnerability of attacks. There are two main types of MITM attacks as shown in Figure 7.4.

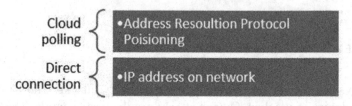

FIGURE 7.4   Types of attacks.

When IoT devices are connected with each other via Bluetooth, MITM attack is leveraged. As Bluetooth is immense for huge power consumption, BLE is used in IoT devices as it is cheaper and cost-efficient. AES-CCM encryption is used to avoid MTIM attacks. BLE uses three-way pairing methodology. The process of the three-way pairing is shown in Figure 7.5.

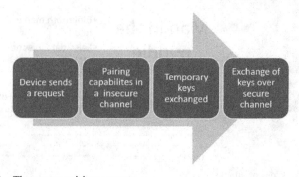

FIGURE 7.5   Three way pairing process.

Fault Data injection is a type of MTIM attack in which IoT-sensored value is changed and prone to suspicion leading to faulty data.

7.3 AI FOR CYBER SECURITY

AI tries to mimic and build the human expertise. AI has various possibilities to provide solutions for cyber security. AI systems are trained to define the threats to identify the defects and to safeguard the information or data. AI systems are tuned perfectly by experts to protect the system from threats. AI-based security systems were developed for cyber security alerts.

7.3.1 MACHINE LEARNING

Machine learning is a subset of AI. ML uses learning methodologies to classify a system. The important learning methods involved in ML are the supervised method of learning and the unsupervised method of learning. Supervised learning is like learning with the help of an expert. A system is trained with all possible inputs and conditions to attain the target output. It uses labeled data or information. Supervised learning methodology again categorizes into two cases like classification and regression. Classification categorizes the input data according to the labeled information. Regression produces a real-time continuous value and acts on conditions. The types of machine learning models are shown in Figure 7.6.

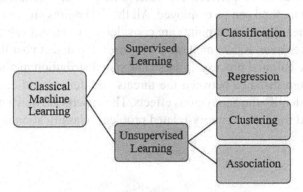

FIGURE 7.6 Classical machine learning methods.

The unsupervised learning method trains the system to describe the information from unlabeled data or information. It will gather all possible combinations of information in a set depending on the inferences made. Unsupervised learning methodology again categorizes into two cases such as clustering and association. Clustering methodology groups data depending on the information provided. An association method builds rules based on the connections between the inputs. ML employs various algorithms for real-world applications. Among these algorithms, most famous one is Naïve Bayes (NB), used for the classification of data based on the Bayesian theorem to supervise all the anonymous activities for cyber attack. NB falls under supervision learning algorithm.[16] Using NB, a system is trained for producing its classes to analyze the cyber attacks and related activities. ML has a variety of algorithms in its pack such as K-Nearest neighbors (KNNs), decision tress, support vector machine (SVM), and so on.

7.3.1.1   K-NEAREST NEIGHBORS (KNNS)

K-NN falls under supervision learning algorithm and is implemented for both classification and regression-type problems. K-NN identifies the optimal dataset for the search space. It classifies the correct dataset depending on the optimal data points and trains the system for a requirement. For K-NN implementation, a training and testing data set is identified. The optimal data point is chosen as K which can be an integer. K point is also referred to as the nearest point for global identification. To find the distance between the other data points, Euclidean distance calculation method can be employed. All the data points are grouped in an array manner. The nearest points are considered as optimal values and the process takes over. K-NN method is very useful to detect yber threats and identify the colonial process. It uses Gaussian distribution method to find the minimum distance between the threats and defects and employs the system to identify the anonymous effects. The experts are down narrowing this method for cyber security-related problems classification is shown in Figure 7.7.

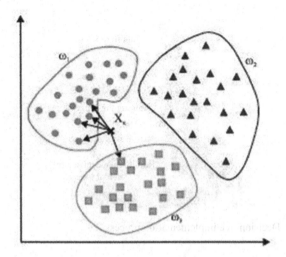

FIGURE 7.7 KNN classification.

7.3.1.2 DECISION TREES

Decision tree (DT) is one of the most powerful methods for classification and identification process. A DT is structured like a tree and its branches, each node denotes specific attributes. It is divided into branches (subsets) that act as rules. DT uses the iteration method to optimize the analysis. It categorizes the event like cyber threat can be categorized as an attack or normal threat. DT trains the system to perform this type of categorization.[17] In cyber security, DT is employed to predict the event. If any anonymous threat is received, the system should identify from the traffic calls whether the received information is categorized under an attack or a normal event. DT uses the data splitting analogy as it falls under the category. All the threats fall under this tree structure and an algorithm is employed for prediction. DT also employs the learning rule technique, which falls under the classification quality, that is, it describes the class of a predicted system. Both methods are employed to predict the cyber threats depending on the anomaly. DT analysis has the potential to support intrusion detection with many challenges in defending the network. The DT has the capability to assist in the analysis of defect information. The implementation process is shown in Figure 7.8

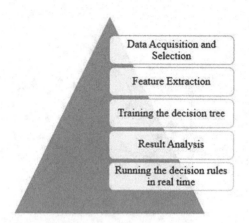

FIGURE 7.8   Decision tree implementation process.

7.3.1.3 SUPPORT VECTOR MACHINES (SVM)

Support vector machine (SVM) employs the supervised learning method for classification problems into pairs. This algorithm is well-suited for text classification tasks, where one usually has contact with a set of values of a maximum of several thousand labeled data samples. The main view of the SVM methodology is to determine an optimal space N (number of points) that single-pointedly finds out the data points. SVM, as already mentioned employs the supervised machine learning technique which finds applications in cyber security in order to determine the threats from training and testing data set of information.

7.3.2 ARTIFICIAL NEURAL NETWORKS

An artificial neural network (ANN) is an optimal methodology to exhibit the human brain inference system. ANN trains the set of inputs with weights and reaches the target output after employing the activation function. The process is repeated till the test value reaches the target value. For this training, various learning rules are employed and continued with weight and bias updation. After calculating the net function, an activating function is employed to determine the target value. The massive advantage of ANNs is that they will adapt their mathematical model as new info is bestowed to them. Whereas alternative mathematical models might become out-of-date as recent methods of traffic and attacks spread. This conjointly means that ANNs are very effective in training the system to determine the cyber threats and to predict malfunctions. Today, the employment of AI in determining cyber defects is an emerging trend now. However, firms with giant networks will enjoy these results, particularly if they are considering or are already implementing IoT devices within their network. AI cyber security also will be helpful in large systems that we tend to encounter in an exceedingly sensible city, and AI can give in no time response times, that is very important for systems corresponding to traffic management or smart home. In addition, several AI cyber security measures sight or stop in-progress attacks instead of forestalling them in the initial place, which suggests that alternative preventive security measures should also exist.

7.4 ARTIFICIAL INTELLIGENCE IN IOT

AI has taken a key role to be combined with the areas of IoT applications in the past 3 years. Most of the startups and other companies that work in the IoT domain are planning to combine AI with their area. Most of the IoT vendors now offer an AI-integrated system with the IoT software platform. This integration has climbed with many opportunities for the people working in the data analytics domain. The data monitoring was done using the IoT platform in the majority of the real-time data acquisition systems. Those data sets can be efficiently interlinked using the AI techniques such as machine learning, deep learning.

Machine learning is one of the AI technologies which inhibits the decision-making scenario by training the data model. Like ANN, the input and output can be prescribed, and using the input layer, hidden layer, and output layer, one can adapt the tuning technique, whereas in machine learning, there are several types of learning as per data prediction type. The Internet of things with AI is blooming fast in comparison with the block chain, edge computing, and other technology. The reason behind this is that IoT is an augmented one and can be enhanced easily using machine learning. This is the one which effectively increases the demand in the business market for AI-based integration techniques in the IoT environment. This makes the success story of AI seem to be an integral part of IoT-based digital ecosystems. The IoT-enabled AI is shown in Figure 7.9.

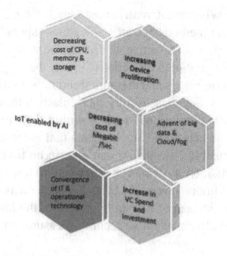

FIGURE 7.9   Drivers for AI growth.

Combining IoT with rapidly fast developing technologies like AI makes the system smarter that will enhance the intelligence which will help in taking decisions without any human intervention. The outcome of this combinatory approach is to accelerate the innovation for boosting the productivity in the organizations. A survey conducted by Bank of America predicted that the growth of the AI market is quickly adhering to the productivity increase of 30%. Simultaneously reducing the manpower labor costs between 18% and 30%. Using the Internet of things in Cyber security had exponential growth by implementing the complex algorithms for better protection in the network systems. When discussing about Cyber security with AI and IoT, it paves a wider way to the cyber attackers. It is indeed to balance both the side of cyber security growth in AI and reduce the cyber-attacking poles. Considering the attacks, not all the AI-based security systems are prone to attacks like intrusion detection, but some of the benefitted AI systems have been tuned against to their self-operation itself. These cases need to be considered for the betterment.

7.4.1   DETECTING VULNERABILITY

To detect the vulnerability in any of the system, machine learning is a good choice for intelligently detecting vulnerabilities which need to be covered. Some people are attacking the vulnerability location and trying to exploit the main system which is running in the backend. Each time, chasing the race with the attacker by manual detection takes more time. This can be reduced by making the automatic detection of capabilities for each day of operation. For each day, the vulnerability count must be minimized which is focusing for a zero-day vulnerability.[18] The main factor for choosing AI for vulnerability detection is the fast fixing of the trap holes. The same job can be also done by the Development team, but the challenge is they used detecting technologies to find each day with each vulnerability; meanwhile, the attacker will use the AI detection and proceed further. Technology for Discovering Vulnerability in Web Interface is shown in Figure 7.10.

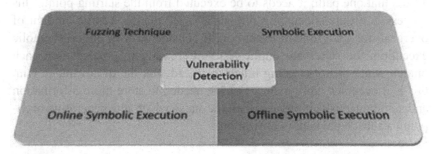

FIGURE 7.10   Technology for discovering vulnerability in web interface.

7.4.2   VULNERABILITY FUZZING TECHNOLOGY

In the area of Cyber security, fuzzing is meant to be an automatic process of discovering hackable bugs in a program by randomly feeding them with automated permutated data, such as comma, separators, braces, etc., creating a stress on the main program to unwantedly create leaks or crashes over the application until it reveals its one of the vulnerabilities. This is a very old technique but for hackers to seek and exploit the vulnerability by bombarding a scratch data into the front end and to result in the issue in the case of zero-day strategy. One has used the fuzzing into the data world to spin up the collaborative work to be spoiled, despite the firewall or password setting, that person can crash the application very soon. In a view, it is not the main goal of fuzzing to crash the program but rather to hijack the program. By putting a bus of data into the program to see what sort of errors they have made on the application.

7.4.3   VULNERABILITY USING SYMBOLIC EXECUTION

This technique is more like the fuzzing technique. The main difference by looking into the fuzzing method is that symbolic execution by using input variables as symbols instead of using a value in searching for vulnerabilities. This technique is categorized into online and offline symbolic execution. The offline method uses only one path for the exploration of path prediction by creating input variables with new values. It is meant to access that one path; it needs to be executed from the starting point. This process is a time-consuming one since it is creating a large amount of overhead for each time of execution. If considered with online symbolic execution, the states are duplicated and create path predicates at each branch which is not creating much overhead. But the disadvantage with this is it occupies more memory space needed to store more information and simultaneously, it processes all the states which it is generating occupying a substantial supply utilization.

7.4.4   DIFFERENT TYPES OF INPUT ATTACKS

The person when attacking the inputs of an intelligent system which causes the system to be crashed or malfunctioning is said to be an input attack. There are n number of attacking methods that can be categorized into different types depending upon the strategy they used for hacking.[19] Like an automated transport system which includes UAV, an automated self-driving car can be hacked by automatically including the physical stop symbol to create a mess in the driving system intelligence. In consideration with these types of input attacks, need not compromise the security system by changing the complexity of the algorithm or adopting different algorithms. But the set of inputs which is malfunctioned for the output needs to be altered. Input attack categorization is shown in Figure 7.11.

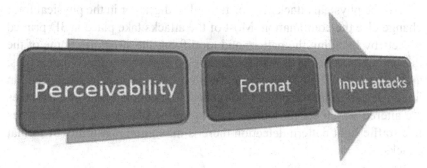

FIGURE 7.11 Input attacks categorization.

Most of the input attacks are like hidden vision from the human eye. Inserting some small image or symbol in the front end also misleads the intelligence by incorporating the algorithm which has been used. Looking into the differentiation of input attacks, it can be classified as perceivable and format type. All these perceivable attacks are visible to the human eye.[20] Some hidden agenda will be lying behind the applications which is an example narrated before like a self-driving car stopped suddenly by inserting a physical stop image into the automated data execution. And noise inclusion in the data comes under the data adding in the system are peculiar types of attacks that might not be noticed by the intelligence. Different forms of input attacks are shown in Figure 7.12.

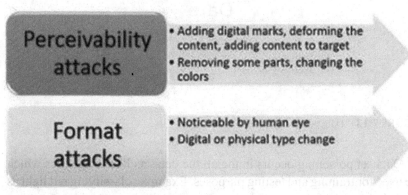

FIGURE 7.12 Different forms of input attacks.

All the physical attacks can be termed as digital or in the physical form change else the combination. Most of the attacks take place in 3D printed objects by blending the pattern and modifying the structure. Some of the digital inputs attacked are video recordings, audio recordings, image files, and video files. In AI-related applications, the format attacks occur in image detection for the prediction-related application. Some of the images are altered which causes malfunctioning in the prediction. In other types like traffic light pattern detection there is maximum possibility of format attacks.

7.4.5   POISONING OF DATA

The training dataset seems to be the most important part of the AI System. This is also attacked by hackers using different types of poisoning models like dataset poisoning, algorithm poisoning, and model poisoning. Data poisoning types are shown in Figure 7.13.

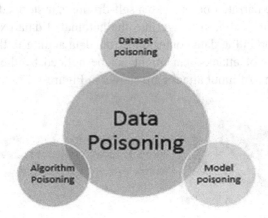

FIGURE 7.13   Data poisoning types.

Data set poisoning occurs in the all the datasets like image files which are used for training and testing purposes. Example: classifying red light as green light, classifying apple image as tomato image, etc., the percentage level of data set poisoning in the malfunctioning is almost 50%. Algorithm poisoning happens in the type of algorithm adopted for detection in

AI systems. Example: federated learning in the G board of Google. The effect of algorithm poisoning is harmful so needs to address soon. Model poisoning can be sorted out by revamping the inputs, data sets.

KEYWORDS

- artificial intelligence
- internet of things (IoT)
- cyber security
- network
- online

REFERENCES

1. Evans, D. The Internet of Things: How the Next Evolution of the Internet is Changing Everything. Cisco Internet Business Solutions Group: Cisco, 2011.
2. Rouse, M. What is IoT (Internet of Things) and How Does it Work? IoT Agenda, TechTarget [Online]. http://www.inter netof thing sagen da.techt arget .com/ defin ition /Inter net-of-Thing s-IoT (accessed Feb 11, 2020).
3. Linthicum, D. App Nirvana: When the Internet of Things Meets the API Economy [Online]. https://techb eacon .com/app-dev-testi ng/app-nirva na-wheninter net-thing s-meets -api-econo my (accessed Nov 15, 2019).
4. Lu, Y.; Xu, L. D. Internet of Things (IoT) Cybersecurity Research: A Review of Current Research Topics. *IEEE Int. Things J.* **2019,** *6* (2), 2103–2115.
5. Vorakulpipat, C.; Rattanalerdnusorn, E.; Thaenkaew, P.; Hai, H. D. In *Recent Challenges, Trends, and Concerns Related to IoT Security: An Evolutionary study,* 2018 20th International Conference on Advanced Communication Technology (ICACT); Chuncheon-si Gangwon-do, Korea (South), 2018; pp 405–410.
6. Lakhani A. The Role of Artificial Intelligence in IoT and OT Security [Online]. https ://www.csoon line.com/artic le/33178 36/the-role-of-artificial -intel ligence-in-iot-and-ot-secur ity.html (accessed Feb 11, 2020).
7. Pendse A. Transforming Cybersecurity with AI and ML: View [Online]. https :// ciso.econo micti mes.india times .com/news/trans formi ng-cyber secur itywith-ai-and-ml/67899 197 (accessed Feb 12, 2020).
8. Meneghello, F.; Calore, M.; Zucchetto, D.; Polese, M.; Zanella, A. IoT: Internet of Threats? A Survey of Practical Security Vulnerabilities in Real IoT Devices. *IEEE Int. Things J.* **2019,** *6* (5), 8182–8201.

9. Roopak, M.; Yun, T. G.; Chambers, J. In *Models Deep Learning, for Cyber Security in IoT Networks*, IEEE 9th Annual Computing and Communication Workshop and Conference (CCWC), Las Vegas, NV, USA, 2019; vol *2019*, pp 0452–0457.

10. Cañedo, J.; Skjellum, A. In *Using Machine Learning to Secure IoT Systems*, 2016 14th Annual Conference on Privacy, Security and Trust (PST); Auckland, 2016; pp 219–222. https ://doi.org/10.1109/PST.2016.79069 30.

11. Farivar, F.; Haghighi, M. S.; Jolfaei, A.; Alazab, M. Artificial Intelligence for Detection, Estimation, and Compensation of Malicious Attacks in Nonlinear Cyber-physical Systems and Industrial IoT. *IEEE Trans. Ind. Inf.* **2020,** *16* (4), 2716–2725. https ://doi.org/10.1109/TII.2019.29564 74.

12. Mishra, S.; Sagban, R.; Yakoob, A.; Gandhi, N. Swarm Intelligence in Anomaly Detection Systems: An Overview. *Int. J. Comput. Appl.* **2021,** *43* (2), 109–118.

13. Raoof, S. S.; Jabbar, M. A.; Tiwari, S. 1 Foundations of Deep Learning and Its Applications to Health Informatics. Deep Learning in Biomedical and Health Informatics: Current Applications and Possibilities, 2021.

14. Gaurav, D.; Rodriguez, F. O.; Tiwari, S.; Jabbar, M. A. Review of Machine Learning Approach for Drug Development Process. In Deep Learning in Biomedical and Health Informatics; CRC Press, 2021; pp 53–77.

15. Pandey, S. R.; Hicks, D.; Goyal, A.; Gaurav, D.; Tiwari, S. M. Mobile Notification System for Blood Pressure and Heartbeat Anomaly Detection. *J. Web Eng.* **2020,** *19* (5-6), 747–773.

16. Roopak, M.; Yun, T. G.; Chambers, J. In *Models Deep Learning, for Cyber Security in IoT Networks*, IEEE 9th Annual Computing and Communication Workshop and Conference (CCWC), 2019.

17. Kuzlu, M.; Fair, C.; Guler, O. Role of Artificial Intelligence in the Internet of Things (IoT) Cybersecurity. *Discover Int. Things* **2021,** *1* (1), 1–14.

18. Jurn, J.; Kim, T.; Kim, H. An Automated Vulnerability Detection and Remediation Method for Software Security. *Sustainability* **2018,** *10*, 1652. https://doi.org/10.3390/su100 51652

19. Comiter, M. Attacking Artificial Intelligence. Belfer Center for Science and International Affairs, Belfer Center for Science and International Affairs [Online]. http://www.belfe rcent er.org/sites /defau lt/files /2019-08/Attac kingA I/Attac kingA I.pdf (accessed Aug 25, 2019).

PART III
Applications of Cybersecurity Techniques for Web Applications

CHAPTER 8

ANALYSIS OF ADVANCE MANUAL DETECTION AND ROBUST PREVENTION OF CROSS-SITE SCRIPTING IN WEB-BASED SERVICES

SMIT SAWANT,[1] GAURAV CHOUDHARY,[2]
SHISHIR KUMAR SHANDILYA,[1] LOKESH GIRIPUNJE,[1] and
VIKAS SIHAG[3]

[1]School of Computer Science and Engineering (SCSE),
VIT Bhopal University, Bhopal, Madhya Pradesh, India

[2]Department of Applied Mathematics and Computer Science,
Technical University of Denmark (DTU), Denmark

[3]Department of Cyber Security, Sardar Patel University of Police,
Jodhpur, Rajasthan, India

ABSTRACT

With the increase in internet users, the number of intruders and attackers is also increasing; this leads to security issues associated with web applications. These web applications are prone to many vulnerabilities due to a lack of secure coding practices by web developers, which in turn risks the privacy and confidential data of end-users using such web applications. Cross-site scripting (XSS) vulnerabilities are one of the most common bugs that affect most modern web applications. The nature of cross-site

The Fusion of Artificial Intelligence and Soft Computing Techniques for Cybersecurity.
M. A. Jabbar, Sanju Tiwari, Subhendu Kumar Pani, & Stephen Huang (Eds.)

scripting has been always evolving and its risk impact on web applications varies from medium severity to critical severity. The traditional defense mechanism could not cope with various bypass techniques to trigger XSS even if there is some kind of protection mechanism to prevent it, for reference the Cuneiform-alphabet based XSS payload bypasses Cloudflare firewall protection mechanism against cross-site scripting attacks. In this chapter, We have analyzed the major concerns for defense mechanisms against cross-site scripting attacks and came up with some robust security solutions which can be integrated with the traditional cross-site scripting defense methodologies to prevent all kinds of cross-site scripting attacks. For finding cross-site scripting bugs, manual detection techniques are robust as it locates vulnerabilities in the website's dark corners while automatic vulnerability scanners have high false-positive and false-negative rates. The overall security of any web application can be enhanced with proposed non-responsive or non-dynamic search boxes which are implemented to prevent cross-site scripting attacks. The combined use of website security plugins with firewalls increases the robustness of web applications. Technique to discover Reflected cross-site scripting (XSS) attacks in URL paths and use of xsshunter, an online platform to find blind cross-site scripting attacks also boosts web applications against all types of cross-site scripting (XSS) attacks.

8.1 INTRODUCTION

Cross-site scripting is a weakness in web applications that can be exploited by a threat actor to gain access to login sessions, open redirects, and so on. Cross-site scripting flaws have been around since a long time, when the World Wide Web was just getting started (Web).[1] The bubble days of old browsers like Netscape, Yahoo, and the obnoxious blink tag were a time when e-commerce started to take off. Website developers do not pay attention to the security aspect totally before deploying their websites and servers in public. If these websites are penetrated, lots of confidential data can be put in danger. According to the research study, almost 40% of all cyber attacks were performed by exploiting the cross-site scripting vulnerabilities. Here is the statistical data of Open Web Application Security Project (OWASP) Top 10 security vulnerabilities for the year 2017 as shown in Figure 8.1.[3]

As we can see in the above prescribed figure that in the year 2017 the percentage of cross-site scripting attacks were about 77% which is really a very big number in terms of cyber attacks. The frequency of cross-site scripting attacks has subsequently increased in large numbers over these years and the risk impact these vulnerabilities can cause is very high. Moreover, the testing for detecting cross-site scripting vulnerabilities is still done in the classical orthodox way, thus many vulnerable sections get overlooked during such penetration testing. Most website moderators rely on old-school defense mechanisms like imposing Content Security Policy (CSP) header, blacklisting approach, escaping characters, encoding techniques, etc., but these defense solutions are not a standalone or robust technique to prevent cross-site scripting (XSS) attacks. Therefore, the website developers need to adapt according to the changing variations of cross-site scripting attacks and devise a robust solution in order to prevent these attacks.

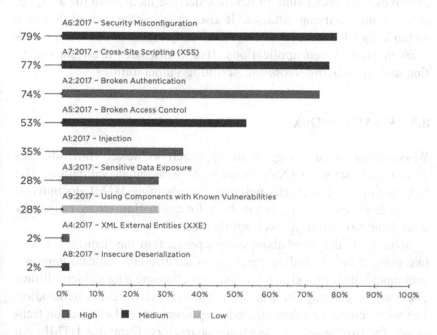

FIGURE 8.1   OWASP top 10 vulnerabilities.[3]

Source: Reprinted with permission from Ref. [3]

8.2   PROBLEM STATEMENT AND OUR CONTRIBUTION

This various cross-site scripting vulnerabilities is so common nowadays that almost 60% bug reports are of cross-site scripting (XSS) on various bug bounty platforms, such as Hacker one, Bugcrowd, Open bug bounty, moreover, the payouts range is also very high as its severity ranges from medium to critical. The main fault is that developers do not practice secure coding methodology. There are various existing works about the detection of Reflected XSS and Stored XSS such as static detection technique, server-side detection techniques. Moreover, there are various client–side and server-side defense mechanisms like nonespaces using randomization, protecting cookies using cryptography, etc., but there are not many solutions on how to prevent DOM XSS and Blind XSS. Moreover, there are various automated mechanisms for detecting cross-site scripting vulnerabilities but we cannot completely rely on them because of high percentage of false-positives. This paper aims to devise a defense mechanism for all types of cross-site scripting attacks. It also provides the manual detection methodology for detecting cross-site scripting areas in various dark areas in modern web applications. It also discussed the technique to find and report blind cross-site scripting vulnerabilities.

8.3   RELATED WORK

Wassermann et al.[19] suggest an approach to detect vulnerabilities of cross-site scripting (XSS) because of weak input validation. The proposed algorithm checks untrusted scrips using HTML documents. Layout engines behavior is the base for evaluating errors caused by weak input validation for web applications.

John et al.[7] discussed about various prevention mechanisms for XSS like using efficient coding practices to sanitize and validate improper data input which provides robust way to eliminate data vulnerabilities. There are several ways by which input sanitization could be handled. The most common methods include replacement and elimination techniques for the search of blacklisted characters. Escaping HTML: All the user input must be HTML escaped. A number of scripting languages, such as jsp, asp, and php support this feature by providing well-defined functions. The approach suggested by Endler and David[4] enables a developer to access "edit" page before "action" page by creating a unique signature based on CGI Script. This mechanism helps in avoiding XSS attacks. Shar et al.[16] suggested a mechanism to escape untrusted reference data in HTML document by OWASP's rules.

Johns et al.[8] studied that the similarity matrix generated using incoming information and outgoing JavaScript can easily detect reflected XSS only for script code. Gupta et al.[6] propose a combined mechanism of browser-embedded script along with IDS which detects malicious JavaScript. The architecture suggested by Gundy et al.[18] defeats XSS attacks in both stored as well as reflected scenarios by returning untrusted user input to the victim.

In Mohammadi et al.[12] cookie misuse is prevented by triple DES encryption of cookie date. Kour et al. [11] discuss about server-side technique which generated hash code for cookie name attributes, and by sending this value to the browser it reduces XSS attacks.

Gupta et al.[6] mentioned about the pros and cons of server-side and client–side approach to avoid XSS attacks. Client–side approaches cause poor surfing experience while server-side approaches increase system overhead.

8.4   CROSS-SITE SCRIPTING (XSS)

Cross-site scripting (XSS) is one of the most popular web application vulnerabilities that allows an attacker to inject malicious client code and alter the interactions that users have with web applications. This vulnerability can be abused by threat actors to hijack user's login sessions by accessing HTTP cookies, deface websites, bypassing cross-site request forgery, open redirects, etc. Cross-site scripting vulnerability has been on the list of Open Web Application Security Project (OWASP)' top 10 web application security risks for decades. The severity of cross-site scripting vulnerability is from medium to critical depending on the logical functionality of web applications.

Cross-site scripting can be majorly categorized into three types – reflected, stored, and DOM. There is another type – blind xss which has gained popularity these days. Moreover, the severity of blind xss ranges from high to critical.

1. **Reflected XSS:** It occurs when a web application receives data in an HTTP request and includes it in an untrusted manner in its immediate response. The most basic entry point for this form is search boxes on websites, where the user's feedback is mirrored in the website's response.

2. **Stored XSS:** Persistent cross-site scripting is another name for it. It happens when a web application accepts user-supplied data in an HTTP request and includes it in an insecure manner in the subsequent response (i.e., it is essentially stored on the server-side). Malicious code inserted in blog comment sections is a simple example.

3. **DOM XSS:** It occurs when client-side code explicitly modifies DOM content or when client-side browser JavaScript processes data in an unsafe manner and then directly writes back to DOM content.

4. **Blind XSS:** This type is caused due to the same reasons just like the other three types but the triggering action differs from the rest of the types. It is coined as blind because the malicious script gets saved on the web server but gets executed in another part of an application or completely in the other application outside of the current domain. The entry points for this type are contact and feedback pages.

There are various pre-existing defense methodologies for preventing cross-site scripting attacks. These defense mechanisms can be client-side protection or server-side protection mechanism. Some of these defense solutions are the following:

1. **Blacklisting approach:** In this technique, the malicious words are listed which need to be blocked in order to prevent XSS attacks. This technique is not feasible because the threat attacker can easily bypass this defense mechanism. Example: Let us say suppose the developer has a blacklisted script word, then the attacker can try various combinations like ScripT or SCript or simply encode it to base64, that is, c2NyaXB0 in order to bypass the blacklist.

2. **White-listing approach:** This approach is opposite to the blacklisting approach. In this approach, the benign characters are only allowed as input from the user thus eliminating the risk of XSS attacks.

3. **Encoding harmful characters:** In this approach, harmful XSS characters like: ', (,), >, <, =, etc., are encoded to Unicode UTF-8 characters or any other encoding in order to safely parse by the server and thus eliminating the risk of cross-site scripting attacks.

4. **Filtering non-benign characters:** This technique is more efficient, as it directly eliminates the malicious characters from the attacker's input. Example: If the attacker's malicious input contains something like this ¡script¿, this approach will filter out the angular brackets and thus only the simple word script is passed to the server.

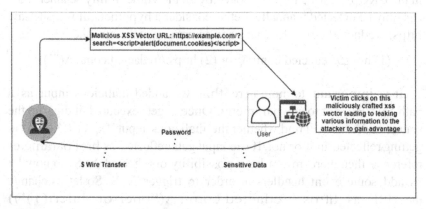

FIGURE 8.2 A generalized demonstration of XSS attack.

8.5 PROPOSED SOLUTION

This dissertation aims to provide techniques for the manual detection of cross-site scripting attacks in modern web applications. It also proposes robust prevention solutions for mitigating the risks associated with different cross-site scripting attacks. The defense mechanism collaborates with the old prevention techniques as well as the newly proposed techniques which will overall help in securing the modern web application.

8.5.1 METHODOLOGIES

Manual cross-site scripting detection techniques are more preferred by bug hunters due to high false-positive and false-negative rates of automated vulnerability scanners. There are various tools like dalfox[1] which scans for cross-site scripting vulnerabilities. This tool tests different XSS payloads on the parameters of various URLs. Though the false-positive rate of this tool is negligible, it still misses various sections for testing XSS in web applications just like XSS getting triggered on the reflected URL path section.

In this section, we will take a look at manual detection of reflected XSS in the URL path, which is overlooked by all the vulnerability scanners as it can only be detected manually. Let us consider a hypothetical website say: https://redacted.com.Now let's add something malicious like:

(1) https://redacted.com/"\¿ or (2) https://redacted.com/(A("'"))

It is important to note here that we added malicious input as a path to URL, not to any parameter. Once it gets executed, it digs up the source code and will find whether the malicious input "\¿ or (A("'")) is getting reflected in it or not. If the input gets reflected as href parameter reference then there might be a possibility of reflected XSS, so now let us add some event handlers in order to trigger XSS. So let us change the URL as https://redacted.com/(A("onerror='alert(1)'")) where onerror='alert(1)' is the event handler; if the web application filters out some of the input, we can change the event handler accordingly like if the alert(1) gets filtered out or the web application firewall blocks the query, then we can try alert '1', it will trigger reflected XSS on web application.

Blind cross-site scripting cannot be detected by automatic scanners. It requires manual inspection as well as automatic platform like xsshunter[2] to generate the report of triggered blind XSS. Most of the contact and feedback pages are vulnerable to blind XSS attacks. Let us see the practical demonstration of detecting blind XSS on a contact page. The xsshunter platform provides different XSS payloads according to different scenarios. Once we decide on the payload to be used for a particular contact form, just fill in all the fields with the payload and submit the form. If the payload gets triggered on the admin or moderator side, the xsshunter platform will notify you directly through email and generate a blind XSS report for the corresponding. The automatic vulnerability scanners are not able to detect such sections where there is a possibility of blind XSS.

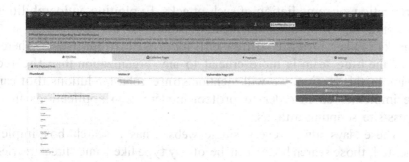

FIGURE 8.3 Blind XSS triggered.

FIGURE 8.4 xsshunter report.

FIGURE 8.5 xsshunter report.

8.5.2 IMPLEMENTATION

Once we are familiar with the different methodologies for detecting cross-site scripting attacks, then comes the most vital part which is the prevention or remediation of such attacks. Exploiting vulnerabilities is not a big deal, but implementing solutions for preventing such attacks is like "Drastic times call for drastic measures." Therefore, the blue team is more crucial to any organization than the red team. In this section, we will mention some robust solutions that can be implemented in order to protect modern web applications from cross-site scripting attacks.

These days almost every single website has a search box implemented, those search boxes can be of any type like some direct queries from the database, or it can be from Google search, or just a normal search box feature provided by content management system (CMS). We all notice that whenever we search any content through these search boxes it will provide us with something similar to a Search result for "XYZ" or No results found for "XYZ," the response of most of these search boxes will be identical to these. The main focus over here is the word "XYZ" which is a user-supplied input that gets reflected on the website, if the website is secure then even if malicious javascript user input is supplied through search box, it will not trigger the reflected XSS. These prevention mechanisms consist of techniques such as escaping harmful characters, sanitizing user inputs, etc., which we have seen in the background section. We devised a search box which is non-dynamic or non-responsive in nature to prevent particularly reflected cross-site scripting attacks. Let us suppose we are searching for a particular word let us say "HELP" through our newly devised non-responsive search box. The non-responsive search box will search the word "HELP" through the whole website as well as a database. If the word does not exist, it will not show any response by reloading the website stating No result found for HELP, so we can infer through this that the user input will not get reflected on the website due to the non-responsive nature of the search box. As we can see in the image below when we searched the word "HELP" in the non-responsive search box it did not show any drop-down menu as the word is not present on the website nor in the database, also it neither reloaded the website to indicate that the word is not present, that is, the user input did not get reflected on the website, thus preventing the possibility of reflected cross-site scripting.

FIGURE 8.6 Non-responsive search box.

Modern WordPress websites are powered with a WordPress security plugin named Wordpfence which includes an inbuilt firewall along with a malware scanner and a vulnerability scanner. So when threat actors provide malicious javascript code in input field to trigger cross-site scripting attacks, Wordfence automatically blocks the request and bans the corresponding IP from which the request was sent to the server. Similar kind of defense mechanism can be implemented for modern web applications, there are various web application firewalls, such as Cloudflare, A kamai, which when organized with a proper rule set for server traffic filtering as well as blocking malicious user requests which helps in preventing cross-site scripting attacks.

FIGURE 8.7 WAF blocking malicious XSS request.

8.6   RESULTS AND DISCUSSIONS

The manual detection techniques discussed above will help the red team of any organization to analyze various vulnerable areas which are prone to cross-site scripting attacks but goes undetected through automatic vulnerability scanners and accordingly convey to the website developer for secure coding practice. The xsshunter platform serves as a great utility for detecting critical blind cross-site scripting vulnerabilities which are not known to website developers as well as website admin or moderator. The non-responsive or non-dynamic search box is the vital concept that various website developers can adopt in order to prevent reflected cross-site scripting attacks. The overall performance metrics associated with this robust solution can be accurately calculated when this type of search box gets implemented in the real-time scenario. It is crucial that the implementation of such search box should be backed up with the traditional defense mechanism to protect cross-site scripting attacks, as there might be a scenario where the threat actor can search a word which is present in the website or in the database, then the search query will be successful for the request, but after getting the result, the threat actor can change the benign script to a malicious script to trigger reflected cross-site scripting attack. The website security plugin like Wordfence should be inbuilt into the website to prevent malicious actors to exploit XSS vulnerabilities by directly blocking them, thus securing the web application. Though Wordfence plugin is developed for WordPress websites, alternate solutions of web application firewalls, such as Cloudflare, Akamai, FortiWeb will efficiently protect the modern web application just like Wordfence plugin.

The proposed system model can be implemented in different information technology sectors, such as financial, banking, insurance, health sectors where it is crucial to protect sensitive and confidential information, such as Personally Identifiable Information (PII), Protected Health Information (PHI), company data, login credentials, and lot more. This proposed system model will help in solving most of the existing problems in traditional defense mechanism for preventing various cross-site scripting (XSS) attacks. The implementation idea of non-responsive or non-dynamic search box concept was an unique way to tackle the problem of cross-site scripting (XSS) attacks. We have also emphasized on the importance of manual detection techniques for finding cross-site scripting vulnerabilities in modern web applications, so as to be more secure and not fall into the mirage of false-positives and false-negatives generated by automatic vulnerability scanners. We also discussed a methodology for finding reflected cross-site scripting vulnerabilities in URL path. The online xsshunter platform will help the web application penetration testers and bug hunters discover rare blind cross-site scripting vulnerabilities in a more efficient and simpler manner. We have also put forth the significance a n d the benefits of implementing security plugins like Wordfence in modern web applications. The web application firewalls, such as Cloudflare, Akamai, FortiWeb, can prove a great alternative to these security plugins but the defense solution will be more robust if they both function mutually. As it is rightly said that no organization is 100% secure due to t h e constant evolving nature of cyber attacks, there is always a scope for improvement for every cyber solution for preventing various cross-site scripting attacks. All these added measures will help improving the conventional methodologies by a significant percentage.

8.7 CONCLUSION

The cross-site scripting is the major issue in the security of web applications. It is necessary to build cyber defense solutions that can prevent cross-site scripting attacks and mitigate the risks associated with them. The crucial challenge was to integrate the new proposed defense solutions with the existing traditional defense mechanisms to work mutually in defending cross-site scripting attacks. Moreover, developers do not focus on secure coding practices thus exposing a bunch of vulnerabilities in modern web applications to the threat actors. The reason behind the poor efficiency of these defense solutions is that they are not updated according to the new and evolving nature of cross-site scripting attacks and their different payloads. The threat actors always discover new techniques to bypass these defense mechanisms and exploit the cross-site scripting vulnerabilities. In this work, the proposed work overcomes the drawbacks of the existing defense mechanisms for preventing cross-site scripting attacks.

The research paper discussed how manual detection techniques can be useful in finding cross-site scripting vulnerabilities in various dark areas of modern web applications where the automatic vulnerability scanner cannot even think of such sections. The technique to discover reflected cross-site scripting (XSS) attacks in URL paths and the online platform xsshunter is used to find blind cross-site scripting attacks in web applications. The new ideology of non-responsive or non-dynamic search boxes is implemented in order to prevent cross-site scripting attacks. These search boxes are not a standalone solution for the prevention of cross-site scripting attacks, as it needs to be backed up with the existing traditional defense mechanisms too. The paper also stressed the importance of incorporating website security plugins like Wordfence in modern web applications, as well as the usability of website security plugins like Wordfence for modern WordPress websites. Though web application firewalls, such as Cloudflare, Akamai, and FortiWeb are great alternatives to the website security plugin, but if these both mechanisms are implemented in modern websites it will boost the overall security of the web applications and thus will robustly defend all kinds of cross-site scripting (XSS) attacks.

KEYWORDS

- cross-site scripting (XSS)
- manual detection
- website
- vulnerabilities
- security
- defense mechanisms

REFERENCES

1. Automatic Cross Site Vulnerability Scanner [Online]. https://github.com/hahwul/dalfox.
2. Automatic Platform to Generate Reports for Triggered Blind xss [Online]. https://xsshunter.com/.
3. Owasp Top 10 Vulnerabilities for the Year 2017 [Online]. https://www.ptsecurity.com/ww-en/analytics/web-application- vulnerabilities-statistics-2019/.

4. Endler, D. The Evolution of Cross Site Scripting Attacks, Technical Report, Technical Report, iDEFENSE Labs, 2002.

5. Gupta, S.; Gupta, B. B. Cross-site Scripting (xss) Attacks and Defense Mechanisms: Classification and State-of-the-art. *Int. J. Syst. Assur. Eng. Manag.* **2017**, *8* (1), 512–530.

6. Gupta, S.; Sharma, L. Exploitation of Cross-Site Scripting (xss) Vulnerability on Real World Web Applications and its Defense. *Int. J. Comput. Appl.* **2012**, *60* (14), 28–33.

7. Mohd Umar, J.; Shah, J. L.; Ahmad, G. I. Web Abuse using Cross site Scripting (xss) Attacks.

8. Johns, M.; Engelmann, B.; Posegga, J. In *Xssds: Server-side Detection of Cross-Site Scripting Attacks*, 2008 Annual Computer Security Applications Conference (ACSAC); IEEE, **2008**, pp 335–344.

9. Kieyzun, A.; Guo, P. J.; Jayaraman, K.; Ernst, M. D. In *Automatic Creation of sql Injection and Cross-site Scripting Attacks*, 2009 IEEE 31st International Conference on Software Engineering; IEEE, **2009**; pp 199–209.

10. Kirda, E.; Jovanovic, N.; Kruegel, C.; Vigna, G. Client-Side Cross-site Scripting Protection. *Comput. Secur.* **2009**, *28* (7), 592–604.

11. Kour, H.; Sen Sharma, L. Tracing Out Cross Site Scripting Vulner- Abilities in Modern Scripts. *Int.J. Adv. Netw. Appl.* **2016**, *7* (5), 2862.

12. Mohammadi, S.; Koohbor, F. Protecting Cookies Against Cross-site Scripting Attacks using Cryptography. In World Scientific and Engineering Academy and Society (WSEAS), 2010.

13. Nadji, Y.; Saxena, P.;Song, D. In *Document Structure In- tegrity: A Robust Basis for Cross-site Scripting Defense*, NDSS; **2009**, vol. 20.

14. Rodr´ıguez, G. E.; Torres, J. G.; Flores, P.; Benavides, D. E. Cross-Site Scripting (xss) Attacks and Mitigation: A Survey. *Comput. Netw.* **2020**, *166*, 106960.

15. Shar, L. K.; Tan, H. B. K. Defending Against Cross-site Scripting Attacks. *Computer* **2011**, *45* (3), 55–62.

16. Shar, L. K.; Tan, H. B. K. Automated Removal of Cross Site Scripting Vulnerabilities in Web Applications. *Inf. Softw. Technolo.* **2012**, *54* (5), 467–478.

17. Louw, M. T.; Venkatakrishnan, V. N. In *Blueprint: Robust Prevention of Cross-site Scripting Attacks for Existing Browsers*, 2009 30th IEEE Symposium on Security and Privacy; IEEE, **2009**; pp 331–346.

18. Gundy, M. V.; Chen, H. Noncespaces: Using Randomization to Defeat Cross-site Scripting Attacks. *Comput. Secur.* **2012**, *31* (4), 612–628.

19. Wassermann, G.; Su, Z. In *Static Detection of Cross-Site Script- ing Vulnerabilities*, 2008 ACM/IEEE 30th International Conference on Software Engineering; IEEE, **2008**; pp 171–180.

CHAPTER 9

SOFT COMPUTING TECHNIQUES FOR CYBER-PHYSICAL SYSTEMS

R.R. RUBIA GANDHI,[1] ABINAYA INBAMANI,[1] N. DIVYA,[1] M. KARTHIK,[1] and E. RAMYA[2]

[1]Sri Ramakrishna Engineering College, Coimbatore, Tamil Nadu, India

[2]Bannari Amman Institute of Technology, Sathyamangalam, India

ABSTRACT

Cyber Physical Systems (CPS) are essential for paving a bridge between the physical objects and the computation elements. The CPS is prone to vulnerabilities and hence its dependability should be overawed by proper modeling of the elements and network. This book chapter gives an indication on the various computing techniques pertaining to CPS. The various spheres including the application domains in smart grids, industrial automation, and intelligent transportation system are discussed. The attributes of the dependability factors are briefed to enhance the security. Soft computing methods are preferred as they are good in solving nonlinear problems and use approximation techniques for self-evolving and improvements. This book chapter describes the soft computing taxonomy with special inclusion to Fuzzy Logic, Artificial Neural Network, and Genetic Algorithm. The various integration technologies with regards to Cyber Physical systems are also described.

The Fusion of Artificial Intelligence and Soft Computing Techniques for Cybersecurity.
M. A. Jabbar, Sanju Tiwari, Subhendu Kumar Pani, & Stephen Huang (Eds.)

9.1   INTRODUCTION

The combination of hardware and software for a specific application is called embedded systems. The embedded system can work either as a standalone or can be connected to the internet. As the application increases, the number of controllers and microprocessors associated with it also increases. When the embedded systems remain connected to the internet, the processor's intensity has been considered. So, for seamless transmission of data from one place to another, the layers of internet have to be considered. These layers will help in establishing the connectivity between various mediums and various operating systems. The security of the data has to be considered when the transmission is done to the end user. The deploy ability, operation, device management, and device discovery are the important factors under consideration. The seamless transfer of information along with the security leads to Cyber Physical Systems (CPS). The cost of processors has declined and the computation speed and its pertaining memory have improved. This leads to a lot of structured, unstructured, and semistructured data. Proper security of the data will in turn help in efficient communication. The initial inventions of wireless sensor networks have led to inventions of higher end communicators like IoT, M2M, V2V, and machine learning techniques. The architecture of the wireless sensor network is shown in Figure 9.1. The important components of WSN are the sensor network, analog circuit, microcontroller, battery, and radio. The sink node, sensor node, cluster node, and routing node are the important nodes in the wireless sensor network domain.

The data from the wireless sensor node is communicated to the concentrator device via ISA, WHART and communicated to the end user with the help of the internet using TCP/IP or Modbus. Security plays a major role during this conversion. In the Gateway solution modbus can be used to connect the concentrator device to the end user. In the TCP/IP solution, the modbus and TCP/IP are used between the concentrator device and the user as shown in Figures 9.2, 9.3, and 9.4. The concentrator devices are given a unique address and data are processed.

The hybrid-based topology solutions have concentrator devices connected to the internet. There is no aggregator point in hybrid solution. The access point solution helps in aggregating the data and proceeding to the end user. The various topology solutions are shown in Figures 9.5 and 9.6. As the advancement of wireless sensor networks proceeds to the next level of embedded systems, the security feature also poses a major role. The physical systems when pertained to the internet have to be identified using IP addresses and secured by using Layer security protocols.

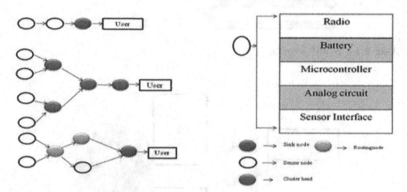

FIGURE 9.1   Components in wireless sensor nodes.

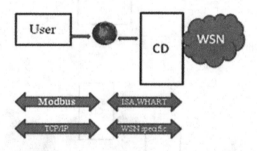

FIGURE 9.2   Front end solution.

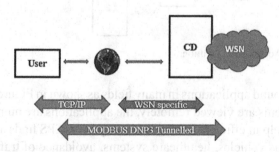

FIGURE 9.3   Gateway solution.

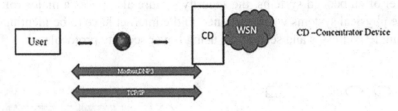

FIGURE 9.4   TCP/IP solution.

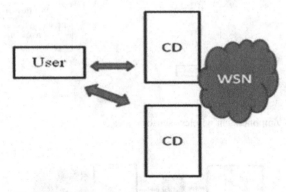

FIGURE 9.5   Hybrid solution.

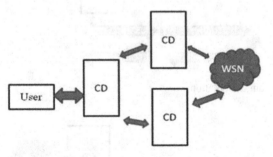

FIGURE 9.6   Access point solution.

CPS has found applications in many fields as shown in Figure 9.7. When physical systems are viewed remotely, the applications are numerous. The smart grids help in efficient management of power. CPS finds applications in autonomous vehicles, healthcare systems, avoidance of traffic congestion, reduction of greenhouse gas emissions, etc.

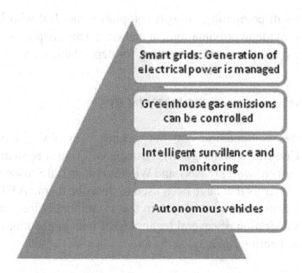

FIGURE 9.7 Applications of CPS.

CPS, when subjected to emergency or unavoidable circumstances, should have the ability to react to it. Dependability and reliability should be given importance and CPS must be robust and stable. Soft computing techniques help in making the CPS stable.

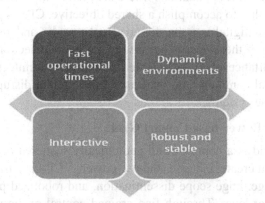

FIGURE 9.8 Features of soft computing techniques in CPSs.

Soft computing techniques can be used for modeling along with its optimization. Accuracy is mandatory in hard computing, whereas robustness is priority in soft computing. As soft computing can deal with imprecise data, it can provide approximations in answers. This chapter briefs about the application domains, soft computing for dependable CPS.

9.2   CLAIM AREAS AND FIDELITY IN CPS

CPS combines computation and networking with physical processes. Networked Control Systems (NCS), Distributed Control Systems (DCS), Sensor Actuator Networks (SAN), and Wireless Industrial Sensor Networks (WISN) are all terms that have been used to describe them.[2] A CPS can be modeled as a distributed control system that is temporally integrated. CPS enables the integration of several technologies with applications across a variety of engineering fields.

9.2.1   CLAIM AREAS OF CPS

CPS has applications in practically every aspect of up-to-date natural life. CPS applications incorporate transportation frameworks, helped medical services, fluid organizations, independent vehicles, keen lattices for utility organizations, and communication, to give some examples. CPS makes an environment in these applications where distinctive implanted frameworks can work together to accomplish a shared objective. CPS is progressively helpless against digital assaults, digital actual assaults, and disappointment circumstances in these spaces because of its interconnection. To emphasize the importance of a dependable and resilient architecture, we will go over several common CPS use cases and different disruptive circumstances in these systems.

1. Electrical Power-based Smart Grid

The power grid is a complex and geologically scattered organization of substances that create, manage, and use power. A CPS is a framework that joins power age, huge scope dissemination, and robotized power control at the customer level. Through fine-grained control of the whole framework, these savvy networks empower higher adaptation to noncritical failure, security, and monetary advantages. Savvy networks can direct constant circulated detecting, estimation, and investigation of electrical force creation and dispersion.[3] These advancements have the advantage of bringing down power interferences and ozone-depleting substance discharges. Notwithstanding these benefits, brilliant matrices are helpless against digital and digital actual attacks, which can bring about foundational harm and a blackout on a worldwide scale. A deliberate cyber-attack on grid operators caused a major outage in Ukraine in December 2015. Around 225,000 people were left without power as a result of this. Smart grid, businesses, and system dependability are both impacted by such assaults. As a result, systems that protect both individual system components and the entire CPS are required.

2. Water Networks

Water networks are fundamental public plans that straightforwardly influence the norm of living. Water networks are very mind boggling, containing an assortment of detecting gadgets, and their intricacy is ceaselessly extending to fulfill the rising requests of enormous urban areas and organizations. Water frameworks are amazingly delicate to a wide scope of dangers. Any digital or digital actual attack can possibly have genuine wellbeing and monetary outcomes. For instance, in 2000, a disappointed worker in Mariachi Shire, Australia, directed a progression of attacks on the SCADA framework working the sewage treatment office, bringing about the spillage of 800,000 liters of crude sewage into public and neighborhoods, exacting critical harm. These examples highlight the need of having a safe and trustworthy infrastructure in place to conduct CPS activities.

3. Industrial Automation

Through a heterogeneous organization engineering of sensors, actuators, and PCs, CPS might empower complete authority over complicated and monstrous mechanical offices.[4] Fusing CPS into the mechanical chain will bring about extraordinary incomes for business and customer adaptability.[28] The fourth modern upset is being proclaimed as the conversion of industry mechanization, PCs, and constant systems administration. From the store network through assembling, stock administration, stockpiling, and exchanging, this can further develop the entire creation cycle. The German government launched the "industry 4.0" project to bridge the gap between seemingly unrelated parts of the supply and manufacturing chain. The development of standards and protocols for communication across the frequently disparate parts of the industrial process is underway. Intelligent solutions in industrial automation will allow the sector to be more responsive to consumer needs. These CPS in industrial automation, on the other hand, are quite susceptible. As a component of a more extensive digital assault in 2013, unfamiliar programmers broke into the control arrangement of a dam in Rye Brook, New York. In 2014, another cyber assault took place in a German steel mill, resulting in massive physical damage. Using spear phishing email, the adversary obtained access to the plant network, causing many components and crucial processes to fail. Intelligent cars and intelligent infrastructure are being integrated into transportation networks.

4. Intelligent Transportation Systems (ITS)

Setting mindful vehicular CPS with cloud help will further develop driver, traveler, and passerby accommodation and security.[5] Such devices will help to alleviate traffic and parking issues in cities. Vehicles may travel together in fleets in a controlled transportation system, and the road infrastructure can be exploited to its full potential. In the event of a crisis, smart transportation will aid in the evacuation of urban residents. While the framework and vehicles required for really keen transportation frameworks are as yet in their beginning phases, the flight area is extensively further developed as far as innovation and organization of correspondence. A disappointment in ITS can have an assortment of natural results, just as time waste and public instability. Such failures can be caused by a variety of security vulnerabilities either by the system designers or by specific ITS components. Ghena et al. of late analyzed the retreat component of a certifiable ITS in Michigan to discover a few security issues. They had the option to distinguish three critical defects in the framework. These incorporated an absence of encryption, an absence of secure validation, and programming weaknesses. The creators took advantage of these blemishes to dispatch an assault on the framework, exhibiting to specialists how a foe might hold onto control of traffic foundation and use it to acquire an out-of-line advantage by lessening wellbeing and causing disturbance.

5. Healthcare

CPS have drawn in a ton of consideration as of late due to their possible uses in medical services.

Wellbeing observing gear like sensors, actuators, and cameras can be joined with digital parts and knowledge in such frameworks. Several CPS designs have recently been developed to improve healthcare facilities.[34]

The WSN-cloud framework is used to showcase a CPS-based secured architecture for healthcare applications. Similarly, Ref. [6] proposes health-CPS model based on a mix of cloud and big data analytics. CPS-based healthcare systems will be able to deliver universal healthcare thanks to developments in IT and AI. As a medical care supplier, CPS gives wellbeing administrations dependent on the wellbeing records or accounts of patients to upgrade therapy and patient consideration.[7] Criminals and cyber dangers have access to personal information stored in healthcare systems. An overall payoff product attack that designated medical care establishments in the United Kingdom, Ukraine, Spain, France, and clinics in the United States is an illustration of such an assault.[8]

9.2.2 DEPENDABILITY IN CPS

The entirety of the above models features the need of making CPS activities heartier and more reliable. Since the applications and administrations presented by a CPS should be guaranteed and dependable in an assortment of conditions (i.e., neighborhood just as worldwide), in this segment, we initially investigate the idea of steadfastness in a more extensive viewpoint prior to zeroing in on it explicitly with regards to CPS. Trustworthiness is a framework quality that incorporates attributes like dependability, accessibility, survivability, wellbeing, viability, and security. It basically takes key standards from an assortment of innovations and joins them into a solitary expression. Constancy is characterized by the International Standards Organization (ISO) as the total word used to portray accessibility execution and it is influencing factors: unwavering quality, execution, viability execution, and upkeep support execution. Constancy is characterized by the International Electro specialized Commission (IEC) as a level of accessibility. Dependability in registering is an attribute of a processing framework that permits the client to believe the assistance it gives. As indicated by the unmistakable specialists nearby, one more definition for constancy is the ability to keep away from administration disappointments that are more incessant and serious than is mediocre. Contingent upon the circumstance, the expression trustworthiness has fluctuated implications. Figure 9.3 portrays the free characteristics of trustworthiness, which include:

- Availability: the ability to provide accurate service;
- Reliability: the capacity to provide proper service on a consistent basis;
- Security: no catastrophic effects on the user(s) or the surroundings;
- Integrity: lack of inappropriate system state changes;
- Confidentiality: absence of unauthorized exposure of information
- Reliability: the ability to be repaired and modified.

In the absolute sense, these characteristics are difficult to define. Treats are unavoidable in actual systems since they are never completely available, dependable, or safe. In the CPS worldview, we by and large inspect two kinds of dangers: arbitrary issues and disappointments, and vital dangers, which are attacks by a foe fully intent on disturbing CPS cycles to the greatest degree conceivable. The formation of a reliable PC framework requires the combination of various strategies and approaches equipped for risk avoidance, danger resilience, danger expulsion, and danger anticipating. The possibility of constancy should be explored as far as dangers to it just as approaches to accomplish it. To be reliable, a framework should have the option to help the accompanying:

- Danger avoidance: how to keep dangers from happening or being presented;
- Threat resistance: how to offer suitable support despite dangers;
- Threat expulsion: how to decrease the number or seriousness of dangers;
- Threat determining: how to conjecture the current number, future rate, and logical outcomes of dangers.

General-reason figuring is much less surprising and reliable than hardware, therefore embedded frameworks should be more reliable as a result of CPS. CPS should be solid and unsurprising to be utilized in significant applications like medical care, airport regulation, and auto wellbeing other variables, like security, should likewise be thought of. The advanced CPS faces vulnerability from both the actual world and digital parts of the framework because of the expanding level of incorporation of new data advances. These blemishes in the CPS can open the framework to an assortment of dangers and perils from assailants, bringing about huge damage. As a result, while developing a dependable and resilient CPS, both digital and actual vulnerability should be considered. CPS vigor alludes to its ability to withstand and be ensured against a known scope of obscure aggravations and boundaries, though its security alludes to its capacity to withstand and be shielded from startling and pernicious events. The CPS is meant to be secure and resilient; thus, these two qualities are pre-event. Despite numerous attempts, creating robust and secure systems is highly expensive, and complete security and robustness is difficult to accomplish. Therefore, survey the framework flexibility (post-occasion), which is the framework capacity to recuperate following the event of problematic occasions. When it comes to the reliability of computing or communication systems, the idea of security comes in useful. Confidentiality, integrity, and availability have all been defined as components of security. Another notion that has expanded significance in the perspective of safety is confidentiality (confidence that information will not be shared without authorization). In terms of the main qualities of dependability, Figure 9.3 illustrates the link between steadiness and safety. The creation of a robust CPS necessitates a thorough knowledge of the effects of cyber assaults. This necessitates a review of CPS cyber infrastructure reliability as well as its capacity to withstand outages. CPS are complex systems with many operational loops operating at various time and spatial scales The dependability of a system component can be used to evaluate its overall reliability. The disappointment likelihood of a framework without excess is higher than the disappointment likelihood of any of its parts. The qualities of a CPS are determined by component attributes as well as the system architecture. Traditional approaches for systems reliability analysis are typically used in CPS reliability and dependability studies, and are some of the contributions to CPS reliability analysis. Far-reaching concentrate on the trustworthiness of CPS is as yet needed to conjecture their dependability and foster methodologies to further develop it. We may use reliability analysis to find issues in telecommunication networks and to calculate the specific redundancy requirements of a network in the design phase, reliability modeling comes before analysis. Later in the plan cycle, when more explicit execution subtleties are known, unwavering quality examination is played out. The development of a model to conjecture the trustworthiness or weakness of a framework dependent on existing information is known as dependability displaying. We can produce trustworthiness measures for a framework utilizing dependability demonstrating. It could be finished utilizing state-based stochastic models like Markov Chains (MC) and Stochastic Petri Nets (SPN) or combinatorial models like Reliability Block Diagram (RBD) and Fault Tree (FT). Combinatorial models give closed-form equations and allow for the description of system dependability in terms of component reliability. They cannot, nonetheless, portray disappointment conditions and asset limitations, which are vital for keeping up with arrangements and characterizing copy measures. Complex repetitive cycles might be addressed utilizing state-based demonstrating, which predicts unwavering quality systematically. They are also useful for forecasting maintenance policies. The potential of a state-space explosion, on the other hand, must be addressed. They cannot, nonetheless, portray disappointment conditions and asset limitations, which are essential for keeping up with strategies and characterizing copy measures Complex excess cycles might be addressed utilizing state-based displaying, which predicts dependability logically. They use a probabilistic approach to model component interactions. Petri Nets (PNs) and SPNs are a sort of BN that might be utilized to all the more viably address the dynamic (transient and circumstances and logical results) conduct of organization parts. They are particularly appropriate for recreating state changes and information stream in confounded frameworks. They support both mathematical and stochastic recreation. A dependability block chart, otherwise called a Dependence Diagram (DD), is a successive and equal plan of squares that portray the probability of disappointment of a framework as far as part unwavering quality (blocks). Just in case there is no less than one series way of working squares over the range of the outline will a framework be addressed by a RBD work. RBDs are intended for frameworks that cannot be fixed and where the request for disappointments is insignificant. The consistent connection between part or sub-framework disappointments is imagined utilizing FT outlines. In FTA, an essential occasion is the top occasion in a shortcoming tree that mirrors a framework occasion of interest that is connected to part disappointments through legitimate entryways. Individual part disappointments lead to the breakdown of a whole framework, as uncovered by FTs and RBDs. Basic parts can be related to the assistance of FTs RBDs and FTs are combinatorial methods because they let us visualize how a number of events might lead to another one happening. At the point when the arrangement of disappointment matters or when fixes are plausible, Markov Chain based displaying is fitting. In both the early and late phases of the plan interaction, FTs and RBDs are utilized to show reliability and foresee accessibility. In subsequent design phases, models based on Markov chains are commonly employed to assess or compare different design options. Traditional analytical methods or simulation tools can be used to examine models created using these or comparable methodologies.

Formal techniques are increasingly being recognized as a helpful tool for building and validating models. Analytical models rely on the complicated system abstraction, simplification, and unrealistic assumptions. This makes them prone to errors, especially in big, complicated systems. In comparison to standard analytic and simulation approaches, formal methods offer a more rigorous manner of analysis. The scope of this article does not include network reliability evaluation, analysis, or modeling. The reader will discover detailed research on reliability analysis. Boolean logic, analytical models, determinism, and crisp categorization are all used in these traditional reasoning and modeling approaches. In the world of modeling, the system (or CPS) is intended to have all of the necessary knowledge to solve a certain issue. Significant data is much of the time available in the real world as tentatively acquired past information and framework conduct dependent on verifiable information yield information. Various arrangements might exist inside a wide scale arrangement space that can accommodate our test by and large. Delicate processing advances are an assortment of versatile PC apparatuses that can adapt to vague information and look for approximations. In digital physical and other complex frameworks, an assortment of delicate processing approaches might be utilized to build framework reliability or model trustworthiness. CPS, dissimilar to sensor organizations, direct proactive tasks characterized by conveyed control circles that get basic criticism from the climate. Moreover, CPS has a wide scope of hub tallies and correspondence abilities. A cross breed framework rises out of this environment of confounded shrewd frameworks, which utilizes fluffy sets, neural organizations, and developmental calculation in different stages or cycles.

9.3 SOFT COMPUTING METHODS FOR CYBER PHYSICAL SYSTEMS

Soft computing methodology is an assortment of computation techniques that incorporate Evolutionary Computation (EC), Artificial Neural Networks (ANN), and Fuzzy Logic (FL) as important individuals.[9] The logical order of soft computing methodology is shown in Figure 9.9. The computational methods are essential and helpful for the perceptive systems.[10] Advanced methodologies like Machine Learning (ML), Probabilistic Reasoning (PR), Chaos Theory, Bayesian Networks (BNs), Rough sets (RS), and Nature inspired algorithms were categorized as the same sets.[11]

These soft computing techniques are getting used to additionally foster the consistency points as relentless quality of convoluted structures. They have in like manner been utilized in exhibiting the steadfast nature of confounded structures and computer networks. They are essential in events when it is rigid to secure a logical model to gauge system reliability. Soft computing methods often act as an additional system for recreation models.[12] They are likewise helpful in tackling complex advancement issues, especially when data are ambiguous or deficient.

CPS is a system which incorporates the communication, computation, and controlling of data service. Cyber physical systems can provide an active control and data service made on the depth incorporation of information, dispersed computation of those data, and dynamic detecting network from the real-world data. The main advantage of the CPS is that it enables people to get the network access wherever the user wants. CPS integrates several technologies to enhance the accessibility. Some are listed below in the figure. By seeing to it, it can be delivered that compared to sensor networks, IoT, and other networks, CPS is a complex interconnected system to access.[20] The complexity can be reduced by reducing the layers in the architecture by which the performance of the network can be efficiently improved. QoS needs to be guaranteed in CPS from end-to-end transmission. QoS of routing is the part where the soft computing can be applied and tested here. Majority of the discussion focus on the research work carried out using the ant colony optimization (ACO) and genetic algorithm (GA).

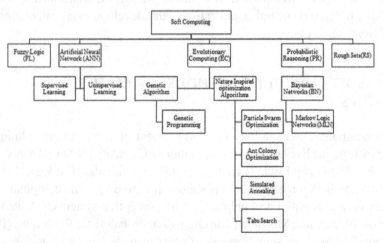

FIGURE 9.9   Soft computing taxonomy.

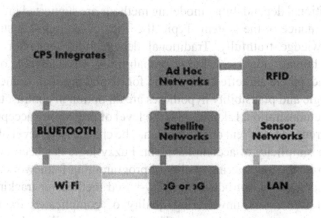

FIGURE 9.10   Integrating technologies in CPS.

9.3.1   FUZZY LOGIC (FL)

Fuzzy set speculation has stood combined to relentless quality model by changing the standard doubts near the determination of a section, that is, twofold state (0 or 1) and probability existence of the constancy.[13] Fuzzy Logic system (FL) was intended to deal with fuzziness utilizing approximate reasoning. It is represented as a kind of believing with the human method for persuasion by means of linguistic variables and characteristics.[11] A system having a fuzzy set can have a level of enrolment for that precise set. Fuzzy arising maps promise to yield using fuzzy logic. This positioning would have the option to be used to derive models.

Fuzzy logic steps into fuzzification, rule estimation, collection of rules, and defuzzification.[14] Fuzzy induction is somewhat easy to execute and discovers broad use in contemporary control framework applications. FL is used to examine the fundamental consistency, safety, fault discovery, security, safety, and risk planning.[15] FL has usually been intensive on the analysis of component reliability where the fuzzy set theory is used to enhance global reliability. Roy et al. have stated the reliability optimization for serial and nonlinear systems with (opposing) consistency and cost goals, using the fuzzy method of multiple goal optimization methods with fuzzy constraints.[16]

Traditional dependability modeling methods are supported information of performance of the system. Typically, it is not likely to get such long-run knowledge truthfully. Traditional dependability behavior likewise includes human judgement. Fuzzy probabilities or conceivable outcomes give an adaptable and effective method for displaying such frameworks.[17] Fuzzy logic and plausibility hypotheses are an option in contrast to probabilistic demonstration. Likelihood is the level of probability accepted from the recurrence of an event or an occasion. The chance is the level of achievability or simplicity of accomplishment. Fuzzy logic has been utilized to demonstrate the unwavering quality of programming frameworks.

Mahapatra et al. introduced a fuzzy-based method for tracking down the ideal framework unwavering quality of complicated frameworks, compelled by a framework cost.[16] Their framework utilizes Intuitionistic Fuzzy Set which is a speculation of fuzzy model hypothesis intended to manage the dubiousness and fuzziness of information. Uncertainties have been utilized to demonstrate human dynamic.[18] Mahapatra et al.[16] compromise some accuracy in dependability streamlining for framework effectiveness. They have utilized fuzzy set hypotheses to deal with loose information and multi-target programming utilizing Fuzzy Nonlinear Programming (FNLP) with fuzzy boundaries. Pandey et al.[19] introduced a fuzzy-based derivation model to foresee programming issues. Their framework needs programming dependability measurements and a system dependent on engineer's capacity development alongside expert's viewpoints.

Fuzzy Logic is a strong and somewhat less complex delicate registering method for order. In many occurrences, assurance of fuzzy participation capacities needed in fuzzy inference systems is made by different procedures. Huang et al.[20] have utilized genetic algorithm to appraise limit upsides of the fuzzy participation capacities, and neural network to gauge fuzzy boundaries for the Bayesian model for dependability investigation. They have utilized fuzzy math with PNs to display dependability with the advantage of expanded adaptability and necessity of a more modest informational index of earlier unwavering quality. Tyagi et al.[21] have determined the dependability of part-based programming frameworks utilizing a versatile neuro-fuzzy surmising framework ANFIS. Fuzzy Logic is likewise utilized related to or to help different strategies for unwavering quality demonstrating improvement or streamlining. Investigation into the use of fuzzy logic in reliable design has focused primarily on the framework dependability analysis. There are a couple of situations where Fuzzy has been utilized for worldwide advancement of unwavering quality.[13]

9.3.2 ARTIFICIAL NEURAL NETWORKS

In view of their organic partners, Artificial Neural Networks (ANN) are hugely equal circulated frameworks for handling data. ANNs can gain from models. They update past gauges considering recently accessible proof.[22] ANNs are organized units that work in correspondence to play out a worldwide assignment. These units will train and test the system boundaries in light of a developing environment.[11] ANNs are utilized in the examination and enhancement of unwavering quality. They are used for boundary assessment for different calculations. The training and expectation capacity make them an imperative apparatus in vigorous way and unwavering quality improvement of CP system. Altiparmk et al. have utilized neural networks to demonstrate the dependability of system.[23] Bhowmik et al.[24] utilized neural networks related to Discrete Wavelet Change to foresee and order transmission line shortcomings. Srivare et al.[25] have demonstrated the gain from existing classifications and anticipate network dependability in an all-terminal organization using ANN. Mora et al.[26] discussed the use of neuro fuzzy classifiers. Caiet al.[27] studied the adequacy of neural network system for taking care of dynamic programming dependability information.

ANNs have been utilized in mix with improvement procedures like GA to appraise beginning qualities for enhancement.[28] Linda et al.[29] utilized the directed ANN-based IDS system for power converter applications. They use Levenberg-Marquardt (LM) calculation and back propagation of error to prepare their structure. Lee et al.[30] proposed a combination of methods with FL rationale regulator for system deployment. The learning ability of neural structure makes them especially appropriate for IDS. Moya et al.[31] have utilized Self Organizing Maps (SOM) for working on the security of sensor information in supervisory frameworks. Kange et al.[32] utilized Deep Learning structure for interruption discovery to work on the security in vehicular networks. Their strategy utilizes high-dimensional CAN bundle information to prepare their profound conviction network which can separate assaulted parcels from ordinary ones dependent on their insights.

9.3.3   EVOLUTIONARY COMPUTATION AND META HEURISTICS

CPS with Genetic Algorithm

All the random search algorithms can be switched into the evolutionary computation techniques. This technique was projected by John Holland., Professor in Michigan University. Based on the genetic evolution given by Mendel and Darwin's Biological evolution theory, GA simulates the bio-genetic and optimistic roles to sort out the solution for heavy population.[34] The main aim of using this approach is to give an optimal solution to a problem using systematic random search. Under evolutionary computation, Genetic algorithms and other types are meant to be special cases. Genetic algorithm is used to stimulate the natural biological genetic behaviors which are necessary for the system enhancing performance.

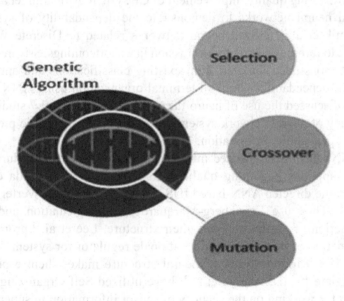

FIGURE 9.11   Optimization flow in GA.

GA is a physical optimization type which uses the optimization process flow of selection, crossover, and mutation. In comparison with the other EC methods, GA has many possible advantages in the subsequent characteristics. It will not fall into the local optimal solution. Foremost point to focus the GA is giving global optimal solutions that can be obtained even if the fitness function is not a continuous one, abnormal, and associated with noise. The next important feature is inherent parallelism, since it is best suited for parallel processing systems in a very large scale. Then it is very flexible to blend with other algorithms to create a new technique which can be used for better random search performance. The main advantages for the optimization of GA are listed here. GA is a very simple, fast, and fault-tolerant algorithm which can be easily adapted for various structured objects.[34]

The GA method does not look for direct variable search, rather it indirectly depends on the variable array of data where those are encoded. Because of the feature, GA can be easily adopted for various structured objects like array, matrix, string, and trees.

The GA method not only searches for a single point, but always tries to do a parallel search for several points in the big population. This parallelism reflects in two ways; as previously mentioned, it is suited for a very parallelism concept in a large-scale sector. It does the search in a wide population space. It looks for several search areas in the solution place and simultaneously exchanges the data or information carried among each involved in the optimization flow.

GA is known for its use of possibility of transferring set of rules for finding out the optimum path in whole population space included.[36] This means that it will search the optimal path among all the individuals involved in the population space, not like the other algorithms which are used to find the path from one of its neighbors. By this characteristic, the chance of falling into local optima can be avoided for sure in GA. Global searching is the biggest challenge with applications like big data analytics, cyber physical systems.

GA has good fault-tolerant capability. In this process, the initial operation of the searching, the population considered for initial operation carries the information or data. This information is not a proper one, that is, irrelevant data deviating from the optimal solution. This can be made as proper information required for optimal solution by using the process of selection, mutation, and crossover. GA has the capability to filter out easily from the vast population solution space using the inherent parallel mechanism. GA conducts random mutation and crossover.[33]

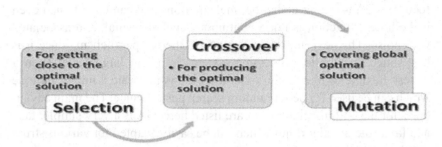

FIGURE 9.12   Random performance operation in GA.

According to the above figure, GA randomly does any of the above-mentioned one process to have a clear path to reach the global optimal solution. These are the important characteristics to choose the Genetic Algorithm for the Cyber physical systems.

9.3.4   ANT COLONY OPTIMIZATION ALGORITHM

Behavior of real ant colonies in nature: In 1990, Dorigo, an Italian researcher, proposed this algorithm. This algorithm mimics the directing performance of life ants. Based on the information exchange and cooperation among the ant colonies, this algorithm finds for the ideal solution.

ACO is known for its dynamic combinatory optimization problem. This is powerful for distributed computing and searching for better solutions. This ACO algorithm can be easily combined with other algorithms and possess good strength in robustness. It has the higher order of flexibility and robustness in a changing environment. It is very flexible enough to be adopted for routing-based problems. From that, it can be taken for the side of network routing problem which is the need for usage in QoS of CPS. The lacuna in this algorithm is delay in searching, lack of initial information, and possibility for stagnation in the move.[33]

As per ACO, each ant involved in the path search dynamically needs to be updated with the Routing Table Item based on the experience and performance of the work given to it. All the ants are united together based on their usual information carried with their smell and hormone secretion called pheromone. If an ant passing path is blocked for a time along, then the path consists of less information element and the respective list item is less enhanced.[37] If any of the paths an ant passes are smooth and provided with more pheromone secretion, then the respective item listed in that path is more enhanced. In this way, the pheromone enhances the routing of the path, but it is time dependent. When surrounding ants stop using this secretion path after a while, it can be inferred that the information was delayed and couldn't be saved for much longer. Because of this move of finding out another new optimal path, even if one path is blocked, it can be adopted for balancing the network load and utilizing such a network for exchanging processes.

In comparison with GA, ACO can also be used as a parallel mechanism. It is a distributive computing method which combines the greedy search algorithm and positive feedback mechanism. It is also a powerful tool for searching optimal solutions. It is known to be parallel computing based on the feature of giving optimal solutions.

This can be implemented in the problem by following the steps listed here.

- Ants—acting as small agents.
- The movement of ant—select the next item for the continuous path movement
- Pheromone—$\Delta \mathrm{T}Ki, j$
- Memory—MK or TabuK
- Another move—use possibility to move ant

9.3.4.1 PROBLEM DESCRIPTION IN CPS

In CPS, there may be presence of several network nodes either wired or wireless sensors or mobile sensors. These are correspondent to the n number of ants in the whole colony included. The system nodes are needed to exchange the data carried to the information center, which can take as food equivalent in the place of information. During the system operation, there might be delay in the network, or any congestion may occur; this needs to be eliminated by the usage of the algorithm.

For analyzing the routing issues, the network can be explained in the form of undirected weigh graph W (n, l) where n is number of nodes, l is the set of bidirectional connection links. The aim of the QoS is to design a multicast tree to solve the routing problems. The factors like delay constraint, bandwidth constraint, Delay jitter constraint, packet loss rate constraint, and cost constraint need to be focused for multicast tree.

9.3.4.2   RULE FOR GA AND ACO

TABLE 9.1   Rules for GA and ACO.

S. No	Rules for genetic algorithm	Rules of ant colony optimization
1.	Genetic Coding—encode the physical nodes	The calculation of transition probability—each ant needs to select the node from i to j
2.	Fitness function—include delay constraint, bandwidth constraint, delay jitter constraint, packet loss rate constraint, and cost constraint	Initial setting of the pheromone—set the initial value of pheromone like $\mathcal{T}s = \mathcal{T}_C + \mathcal{T}_G$
3.	The generating of the population—use rand function to generate the random number of individuals	Updating model of pheromone—local optimal and global optimal path is combined

As is discussed above, for a CPS heterogeneous network, QoS multicast routing issue is considered and by using GA, ACO the following improvements can be obtained. GA implies good searching and gives a global optimal path. Using ACO, positive feedback mechanism and parallelism can be obtained. One more advantage by combining GA and ACO, GAAC is a new algorithm which enhances the multicast routing issue very well.[36] Sample simulation results for the review have been mentioned below, to have a choice of choosing GAAC algorithm for Multicast routing issue in Cyber Physical Systems.

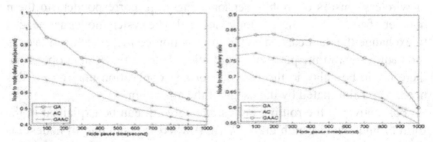

FIGURE 9.13   Simulation figures for GA, ACO, GAAC for node-to-node delay time and node to node to delivery ratio.

Comparing the node-to-node pause time of GA, ACO, and GAAC, GAAC has very short node pause time which will decrease the transmitting delay in exchanging the node information. Comparing the packet delivery ratio of GA, ACO, and GAAC, GAAC is higher. From the above discussion, it is giving a clear picture that usage of CPS along with the soft computing techniques like GAAC (Genetic Ant Colony algorithm) the delay time, cost constraint in the routing networks can be minimized and efficiently improved with faster transmission of information within the information center.

KEYWORDS

- cyber physical systems
- application domains
- ANN
- fuzzy logic
- soft computing
- genetic algorithm

REFERENCES

1. Abinaya, I.; Manivannan, D. Remote Monitoring Using Wireless Sensor Node with IoT. *Res. J. Pharm. Biol. Chem. Sci.* **2015,** *6* (3), 1480–1484.
2. Mitchell, R.; Chen, R. Effect of Intrusion Detection and Response of Reliability of Cyber Physical Systems. *IEEE Trans. Reliab.* **2013,** *62* (1), 199–210.
3. Baheti, R.; Gill, H. Cyber-Physical Systems. The Impact of Control Technology, 2011; vol. *12*, pp. 161–166.
4. Wang, Y.; Vuran, M. C.; Goddard, S. Cyber-Physical Systems in Industrial Process Control. *ACM Sigbed Rev.* **2008,** *5* (1), 12.
5. Wan, J.; Zhang, D.; Zhao, S.; Yang, L.; Lloret, J. Context-Aware Vehicular Xyber-Physical Systems with Cloud Support: Architecture, Challenges, and Solutions. *IEEE Commun. Mag.* **2014,** *52* (8), 106–113.
6. Zhang, Y.; Qiu, M.; Tsai, C. W.; Hassan, M. M.; Alamri, A. Health- cps: Healthcare Cyber-Physical System Assisted by Cloud and Big Data. *IEEE Syst. J.* **2015.**
7. Latif, S.; Qadir, J.; Farooq, S.; Imran, M. A. How 5g Wireless (and Concomitant Technologies) will Revolutionize Healthcare? *Future Int.* **2017,** *9.*
8. Glaser, U.S. Hospitals have been Hit by the Global Ransomware Attack [Online]. https://goo.gl/n4uEk5 (accessed Dec 12, 2017).

9. Divya. N.; Nirmalkumar. A.; A Survey on Tuning of PID Controller for Industrial Process Usingsoft Computing Techniques. *Int. J. Pure Appl. Math.* **2018**, *11* (118), 663–667.

10. Zadeh, L. A. Soft Computing and Fuzzy Logic. *IEEE Software* **1994**, *11* (6), 48–56.

11. Chaturvedi, D. K. In *Soft Computing: Techniques and its Applications in Electrical Engineering*; Springer, 2008; 103, p 11.

12. Ebrahimipour, V.; Asadzadeh, S.; Azadeh, A. An Emotional Learning-based Fuzzy Inference System for Improvement of System Reliability Evaluation in Redundancy Allocation Problem. *Int. J. Adv. Manuf. Technol.* **2013**, 1–16.

13. Ravi, V.; Reddy, P.; Zimmermann, H. J. Fuzzy Global Optimization of Complex System Reliability. *IEEE Trans. Fuzzy Syst.* **2000**, *8* (3), 241–248.

14. Negnevitsky, M. In *Artificial Intelligence: A Guide to Intelligent Systems*; Pearson Education, 2005.

15. Cai, K. Y. System Failure Engineering and Fuzzy Methodology an Introductory Overview. *Fuzzy Sets Syst.* **1996**, *83* (2), 113–133.

16. Mahapatra, G.; Roy, T. Reliability Optimisation of Complex System using Intuitionistic Fuzzy Optimisation Technique. *Int. J. Ind. Syst. Eng.* **2014**, *16* (3), 279–295.

17. Onisawa, T. An Application of Fuzzy Concepts to Modelling of Reliability Analysis. *Fuzzy Sets Syst.* **1990**, *37* (3), 267–286.

18. Atanassov, K. T. Intuitionistic Fuzzy Sets. *Fuzzy Sets Syst.* **1986**, *20* (1), 87–96.

19. Pandey, A. K.; Goyal, N. A Fuzzy Model for Early Software Fault Prediction using Process Maturity and Software Metrics. *Int. J. Electron. Eng.* **2009**, *1* (2), 239–245.

20. Huang, H. Z.; Zuo, M. J.; Sun, Z. Q. Bayesian Reliability Analysis for Fuzzy Lifetime Data. *Fuzzy Sets Syst.* **2006**, *157* (12), 1674–1686.

21. Tyagi, K.; Sharma, A. An Adaptive Neuro Fuzzy Model for Estimating the Reliability of Component-Based Software Systems. *Appl. Comput. Inform.* **2014**, *10* (1), 38–51.

22. Bonissone, P. P. Soft Computing: The Convergence of Emerging Reasoning Technologies. *Soft Comput.* **1997**, *1* (1), 6–18.

23. Altiparmak, F.; Dengiz, B.; Smith, A. E. A General Neural Network Model for Estimating Telecommunications Network Reliability. *IEEE Trans. Reliab.* **2009**, *58* (1), 2–9.

24. Bhowmik, P.; Purkait, P.; Bhattacharya, K. A Novel Wavelet Transform Aided Neural Network Based Transmission Line Fault Analysis Method. *Int. J. Electr. Power Energy Syst.* **2009**, *31* (5), 213–219.

25. Srivaree-ratana, C.; Konak, A.; Smith, A. E. Estimation of all Terminal Network Reliability using an Artificial Neural Network. *Comput. Operat. Res.* **2002**, *29* (7), 849–868.

26. Mora, J.; Carrillo, G.; Perez, L. In *Fault Location in Power Distribution Systems using ANFIS Nets and Current Patterns*, Transmission & Distribution Conference and Exposition: Latin America, 2006; IEEE/PES; IEEE, 2006, pp. 1--6.

27. Cai, K. Y.; Cai, L.; Wang, W. D.; Yu, Z. Y.; Zhang, D. On the Neural Network Approach in Software Reliability Modeling. *J. Syst. Software* **2001**, *58* (1), 47–62.

28. Gao, W.; Morris, T.; Reaves, B.; Richey, D. In *On Scada Control System Command and Response Injection and Intrusion Detection*; eCrime Researchers Summit (eCrime), 2010; IEEE, 2010, pp 1–9.

29. Linda, O.; Vollmer, T.; Manic, M. In *Neural Network-Based Intrusion Detection System for Critical Infrastructures*, International Joint Conference on Neural Networks, 2009; IEEE, 2009, pp 1827–1834.

30. ChangYoon, L.; Tsujimura, Y. Reliability Optimization Design using Hybrid nn-ga with Fuzzy Logic Controller. *IEICE Trans. Fundam. Electron. Commun. Comput. Sci.* **2002,** *85* (2), 432–446.

31. Moya, J. M.; Araujo, A.; Bankovi´c, Z.; De Goyeneche, J. M.; Vallejo, J. C.; Malagón, P.; Villanueva, D.; Fraga, D.; Romero, E.; Blesa, J. Improving Security for Scada Sensor Networks with Reputation Systems and Self-organizing Maps. *Sensors* **2009,** *9* (11), 9380–9397.

32. Kang, M. J.; Kang, J. W. Intrusion Detection System using Deep Neural Network for in-Vehicle Network Security. *PloS One* **2016,** *11* (6).

33. Gandhi, R. R.; Kathirvel, C. In *A Comparative Study of Different Soft Computing Techniques for Hybrid Renewable Energy Systems*, 2021 5th International Conference on Trends in Electronics and Informatics (ICOEI); IEEE, June 2021; pp 1667–1677.

34. Wang, Z.; Jin, Y.; Yang, S.; Han, J.; Lu, J. An Improved Genetic Algorithm for Safety and Availability Checking in Cyber-Physical Systems. *IEEE Access* **2021,** *9*, 56869–56880.

35. Atif, M.; Latif, S.; Ahmad, R.; Kiani, A. K.; Qadir, J.; Baig, A.; ... ; Abbas, W. Soft Computing Techniques for Dependable Cyber-Physical Systems. *IEEE Access* **2019,** *7*, 72030–72049.

36. Gao, Z.; Ren, J.; Wang, C.; Huang, K.; Wang, H.; Liu, Y. A Genetic Ant Colony Algorithm for Routing in CPS Heterogeneous Network. *Int. J. Comput. Appl. Technol.* **2013,** *48* (4), 288–296.

37. Mishra, S.; Sagban, R.; Yakoob, A.; Gandhi, N. Swarm Intelligence in Anomaly Detection Systems: An Overview. *Int. J. Comput. Appl.* **2021,** *43* (2), 109–118.

CHAPTER 10

CYBER SECURITY USING MACHINE LEARNING APPROACHES: A SYSTEMATIC REVIEW

MOHAN KUMAR K N[1] and PAVAN KUMAR E[2]

[1]Department of Computer Science and Engineering, Saividya Institute of Technology, Bengaluru, Karnataka, India

[2]Department of ECE, Saividya Institute of Technology, Bengaluru, Karnataka, India

ABSTRACT

Machine learning in cyber security has recently made headlines. With the widespread increase of population, the need for reliable mechanism to prevent cyberattacks has increased in manifold. In the recent days, there is an increase in cyber security problems in majority of the enterprises across the globe. The reason for cyberattacks problems is not specific but it has become so uncertain. If we take a sample from the population, it should not be a surprise to see a enterprise suffered from cyberattacks irrespective of the technology adopted; for example Ransomware, Spyware, and Trojans are commonly found. So, this situation poses a serious challenge for researchers to find the root cause. It is difficult to accurately predict the future cyberattacks based on the current status because the scenario might not be the same for all the enterprises. Providing an affordable, high quality Cyber security service has become a big challenge. In this sense, cyberattack avoidance has been studied for decades, which is an area with

The Fusion of Artificial Intelligence and Soft Computing Techniques for Cybersecurity.
M. A. Jabbar, Sanju Tiwari, Subhendu Kumar Pani, & Stephen Huang (Eds.)
© 2024 Apple Academic Press, Inc. Co-published with CRC Press (Taylor & Francis)

a steady stream of new work and improvement over time. This article is based on a systematic investigation that aims to bring together prior research on cyberattack prediction and categorization, highlight significant changes in patterns, and propose research direction for future work.

10.1   INTRODUCTION

Preventive care of cyberattack is a concept, in any sort of intelligence, technology or an application of intelligence, that is, cyberattack data from different sources and using predictive human behavior for dispensing information that improves the ability of detecting cyberattacks. Cyberattack is said to be the biggest threat that majority of enterprises face. In the view of researcher's perspective cyberattacks can be defined as any attempt to gain unauthorized personal computer, computing platform, or compute cluster with the intent to cause harm. Cyberattacks are designed to disable, disrupt, destroy, or seize control of computers, and also corrupt, block, wipe, edit, or acquire stored data on them. Cyber security can be disrupted by different attacks from time to time. The functioning of the compute node in the network would become abnormal for various reasons such as financial gain, disruption and revenge, and cyber warfare. Cyberattacks may also set off due to effects of long-term usage of the same technology and one attack may lead many other similar attacks if not taken care of.[1]

Bad actors use cyberattacks to generate chaos, uncertainty, discontent, frustration, or mistrust. They could be acting in this way as a kind of retaliation for actions taken against them. They could be seeking to embarrass the targeted entities or ruin the organizations' reputations. Attacks on government agencies are common, but they can also target financial as well as non-enterprises. Nation-state attackers are behind several of these cyberattacks. Others, known as malevolent hackers, may launch similar assaults in the protest of the recognized entity; the most well-known of these organizations is completely undetectable, a secretive, decentralized group of international dissidents.[2]

A cyberattack is an abnormal situation that affects the normal functioning of the system. Cyberattack is learnt by their indications, which may be caused by dysfunction. Cyberattack can affect enterprise applications not only physically, but also emotionally, as contracting and living with a cyberattack can alter the perspective of the system.

Every year, cybercrime rises dramatically as attackers become more efficient and sophisticated.[3]

Cyberattacks can occur for a variety of reasons and in a variety of formats. Cybercriminals will look to exploit vulnerabilities in an organization's security policies, methods, or technology, according to a common thread. Although an attacker can access an IT system in a variety of methods, the majority of cyberattacks use the same techniques. Among the most popular types of cyberattacks are listed following.

Malware is a type of computer software that can be used to do a variety of damaging tasks.

Certain malware strains are intended to capture everlasting network access, whereas others are developed to spy on the user in order to acquire identities or other important information, and still others are just meant to disrupt the user's life.

Whenever an offender tries to fool an unwary target into giving usernames and passwords, credit card details, proprietary information, and so on, this is known as phishing. Phishing attacks usually take the form of an email appearing to be from a respectable source, such as your bank, the IRS, or perhaps another trustworthy origin. A man-in-the-middle (MITM) attack occurs when an intruder captures communication between the different parties with the intent of spying on the victims, stealing personal details or credentials, or altering the dialogue in a certain manner.

An attacker uses a DDoS (Distributed Denial of Service) assault to overwhelm a target computer with traffic in the aim of disturbing, if not entirely closing it down. Unlike traditional denial-of-service assaults, which are detectable and respondable for most contemporary firewalls, a DDoS attack can leverage several sources of traffic to overload the target.

Structural Query Language (SQL) injection is a specific type of attack that exclusively affects SQL databases. SQL queries are used to search data in SQL databases, and they are usually executed using an HTML form on a website. The attacker may be able to utilize the HTML form to perform queries that generate, read, edit, or delete information from the database if the database privileges are not set up correctly.

A zero-day attack happens when hackers discover weaknesses in popular software programs and systems and then target organizations that use them to exploit the weakness before a fix is issued.

DNS tunneling is a sophisticated attack method that allows hackers to keep access to a target for an extended period of time. Because many firms fail to monitor DNS traffic for malicious activity, attackers can insert or "tunnel" malware inside DNS inquiries (DNS requests sent from the client to the server). The virus is used to create an unbreakable communication channel that is undetected by most firewalls.

Business Email Compromise (BEC): In a BEC attack, the hacker targets specific people, usually workers with bank transaction authorization, and convinces them to transfer the funds to an account owned by the hacker.

Hackers gain entry to a user's computer or mobile device and use that to mine cryptocurrency such as Bitcoin, which is known as crypto jacking. While less well-known than other attack avenues, crypto jacking should not be neglected.

When an unwary victim visits a website that then infects their computer with malware; this is known as a "drive-by-download" attack. It is possible that the website in question is one that the attacker has direct control over or one that has been compromised.

Cross-site scripting attacks, like SQL injections, impact other users who visit the site rather than stealing data from the database. A simple example would be the comments section of a website.

A password attack, as you might have guessed, is a type of cyberattack in which an attacker tries to guess, or "crack," a user's password. There are many different methods for breaking a user's password, and a description of all of them is much beyond the scope of this paper. Eavesdropping, also known as "snooping" or "sniffing," occurs when an attacker looks for unsecured network connections to intercept and access data being sent across the network. For this reason, employees must use a VPN when accessing the company network from an insecure public Wi-Fi hotspot.

Artificial Intelligence (AI) being used to launch sophisticated cyberattacks is a terrifying prospect, given that we have no idea what these attacks will be capable of. The most well-known AI-powered attack to date used AI-powered botnets to mount a huge DDoS attack using slave PCs.

IoT devices are now less secure than most modern operating systems, and hackers are ready to exploit these holes. Because the internet of things, like artificial intelligence, is still a novel concept, we do not know what strategies cybercriminals would use to exploit IoT devices or for what goals.

Efforts are needed to build sustainable preventive care of cyberattacks to target the challenges created by the growing prevalence of enterprise application, with strategies directed to primary, secondary, and tertiary levels of intervention. Decision-makers need information about the type of cyberattack problems, how to address these challenges and how effective these measures have been. Enterprise applications need access to information about their cyberattack to help themselves.[4–10]

The advancements in Machine Learning (ML) have been remarkable; the information provided is better than the existing techniques. Machine Learning enables increased cyberattack analytics, which is important if technology is to improve preventive care in the future. Machine learning models use massive amounts of computational power to establish a set of rules that a human brain would be impossible of processing. The more data given into a machine learning algorithm, the more complicated the rules become, and the more accurate the cyberattack prediction becomes. We need to identify specific use cases where machine learning capabilities can assist us in preventing cyberattacks.

Hence, there is a need to analyze and design an efficient system which can classify attacks. A systematic approach is required to derive knowledge out of cyberattack records which can help to prevent future attacks. Exploration of the existing techniques would give better visibility into the context discussed. IEEE Xplore, Science Direct, SpringerLink, ACM Digital Library, Wiley Online Library, and Google Scholar were utilized to gather information for the reviews.

The structure of the chapter is as follows: Cyberattack prediction techniques are described in Section 2. Datasets which are used in this domain are discussed in Section 3. Features used for cyberattack prediction are introduced in Section 4. Feature selection techniques are illustrated in Section 5. Model performance measures which are frequently used in this domain are explored in Section 6. Future works and conclusion remarks are explained in Sections 7 and 8, respectively.

10.2 CYBERATTACK PREDICTION TECHNIQUES

The techniques used in cyberattack prediction till now can be categorized into parametric and nonparametric. The parametric techniques are Logistic Regression, Linear Discriminant Analysis; they have proved better performance and accuracy in forecasts. Figure 10.1 clearly consolidated from certain causes to find out the procedures that are majorly used.

Logistical regression techniques for cyber-security prediction dominated, with 27.5%, this is because of the nature of health data.[6–10]

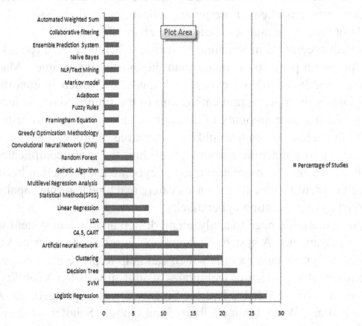

FIGURE 10.1   Prediction techniques.

The nonparametric approaches could deal with the nonlinearity in cyberattack data. When the database is nonlinear, methods like neural networks (NN) and clustering have shown greater prediction accuracy, indicating a better representation of real-world data. Figure 10.1 depicts the number of studies conducted on various forecasting approaches. The second most popular technique (25%) is the Support Vector Machine, which is also the most popular among nonparametric methods. Other NN techniques, such as fuzzy neural networks and neural networks, have produced positive outcomes. At 22.5%, Decision Tree approaches are also popular. Nonparametric models are computationally expensive and cannot be used for real-time prediction without extensive optimization. This is simply one of the reasons to combine statistical and machine learning techniques, that is, combine the capabilities of nonparametric methods to efficiently model nonlinear data. By combining the advantages of many models or by integrated forecasting, which combines the expected values of distinct models, ensemble prediction machine learning models have shown to improve prediction accuracy. In cyberattack prediction, ensemble/combined/hybrid forecasts are often used. The various techniques are employed in different contexts and their employment is listed in Figure 10.1; however given that deep learning and big data are newer breakthroughs, it is somewhat encouraging that the latest technology is now being used in cyberattack prediction.[10–30]

10.3 DATA SETS USED FOR CYBERATTACK PREDICTION

Many of the studies have been conducted on standard database. Other major source of data is KDD-99 which is extensively used (30%); the authors have gathered this data from different sources. Studies are also conducted on synthetic Record to an extent of 10%, 22% of CICIDS 2017 dataset is used to classify different cyberattacks. To achieve optimization studies have been conducted on mentioned data sets as well as on other data sets such as crawled data, six gill data, binomial classification data set, social analysis data, and other log data. It is evident from the studies that certain data set has been used to study the specific attacks. Interesting part is that KDD-99 data is used for the prediction and classification of cyberattacks. Figure 10.2 depicts the count of different data set usage in cyberattack prediction.[5]

TextGAN, FM-GAN, MidiNet, Age-cGAN, CVAE-GAN, SenseGen, WGAN, ACGAN, Pedestian synthesis GAN, HP-GAN, VAE-GAN, WaveGAN, DermGAN, CT-GAN, X2CT-GAN, and D-NET are some synthetic data generation AI algorithms.[31]

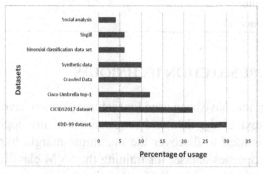

FIGURE 10.2 Plot of data sets used for cyberattack prediction.

10.4   FEATURES USED FOR CYBERATTACK PREDICTION

From Table 10.1, it is clear that professional crimes, corporate rivalry, and action origin are the most used feature in cyberattack prediction (10–38%); these features are commonly associated features along with other features. Professional crimes, corporate rivalry provides a description of the attack state at different phases. From this study it is deciding factor in attack prediction. To achieve precision, features such as action flood,[23] timeliness,[30] and autonomous[23] are also studied (2–10%). Interestingly some of the studies have proved that features such as availability status[19] help in attack prediction. The latest trend is that timeliness[15,24] described in data has been used for prediction analysis. The other commonly studied features that are used in process of attack prediction and prevention are autonomous.[22] Some studies have used non-traditional methods, such as HTTP locale information.[21]

TABLE 10.1   Comparison of the Percentage of Studies on Cyberattack Prediction Based on Features.

Sl. No.	Features	Percentage of usage
1	Professional crimes	38
2	Corporate rivalry	20
4	Action origin-remote (single source)	10
8	Availability	10
5	Action origin-remote (multiple source)	8
6	Action-flood (single source)	5
7	Action-flood (multiple source)	4
9	Timeliness	3
3	Autonomous	2

10.5   FEATURE SELECTION TECHNIQUES

The techniques that have been used for feature selection have been detailed based on context of the data and requirement of the applications. The linear kernel is used to discover the maximum-margin hyper-plan from the Context properties dataset for training the SVM classifier to find the maximum-margin hyper-plan.[13] Construction of several sparse matrices from a temporal aggregate of claims data and the matrices thus created can be viewed as a document-term matrix which can be used for training machine learning models.[14] The commonly used feature selection techniques are Re-sampling and Attribute Selection Techniques on manually collected data.[24] Directly from the data set, rule learners extract rules that explain a pattern or correlations between input features and output class labels.[30] AI-based models are used to develop term document matrix for diagnosis data.[24] Lastly some statistically significant predictor factors are used commonly for feature extraction.[30]

10.6 MEASURES OF PERFORMANCE EVALUATION

The most often utilized performance measurements are root mean square error (RMSE) and mean absolute percentage error (MAPE).[24,25] RMSE with cost (RMSEC) and RMSE variants such as Normalized RMSE (NRMSE) and RMSE with cost (RMSEC) were also employed. Mean Absolute Relative Error (MARE),[17] Equal Coefficient (EC), R square, Mean Square Error (MSE), Mean Relative Error (MRE), Accuracy,[29,30] and Variance Absolute Percentage Error (VAPE) are used in substantial quantities, but not nearly as much as the ones described above. Precision, recall, F1, False Positive rate, sensitivity[24,25,29,30] and all these are the performance measures that are unique to the field of cyberattack forecasting. Because all of the performance indicators add up to a considerable portion of the total, it is difficult to compare approaches between different works.

10.7 AI-POWERED TOOLS

DeepHack, DeepLocker, GyoiThon, EagleEye, Malware-GAN, uriDeep, Deep Exploit, and DeepGenerator are few popular AI-powered tools for offensive cyber operations which are mainly used for the data analysis.[31]

10.8 FUTURE WORKS

Researchers are working effectively in this domain. It is very much essential to understand the future scopes in the prediction of cyberattack such as:

- More studies are essential to find effective predictive models which produce positive outcomes. Several researches have contrasted Artificial Neural Networks to various machine learning approaches. As a result, more studies compare and monitor the effectiveness of Artificial Neural Network approaches with other state-of-the-art approaches that are needed. [13,20,26].
- A regular study is necessary to understand sophisticated defense systems using machine learning systems to protect from adversarial attacks.
- The existing models consider small set of records and features. They do not consider the overall history of attacks with respect to hierarchy of attacks.[17]
- Present models do not predict multiple attack correlation and their hierarchy of occurrences.[24]
- Researchers of this domain need to adopt big data technology to study cyberattack records of large size.[30]
- It is very much essential to investigate Dark Web sources like IRC chat rooms, online market places, etc.
- The goal of classification approaches is to discriminate between accurate and erroneous data, not to extract correct data. As a result, it is an important thing in the future to bring attention to which datasets should be included in inaccurate data.
- Understand the gap between the amount of research done in this field and the amount of specific procedure of suggested technique.
- Unsupervised deep leaning technique with k-means clustering can be used for better results.

10.9   CONCLUSION

A complete assessment of the existing studies on cyberattack prediction for preventive treatment of attacks is examined in this article, with the goal of recording and aggregating the findings. The requirements for a systematic study were first determined. The research needs were stated in such a way that inferences from our findings could be made. There are several prediction and classification strategies for cyberattack prediction that have been identified. We searched the internet for relevant research on the topic, extracted important data, and published a summary of our findings to give an overview of the field, as well as to make one up to date on the demands for cyberattack forecasting and to propose recommendations for future work. The information derived from this research will aid our future efforts to establish a cyberattack prevention strategy.

KEYWORDS

- cyber security
- cyberattacks
- prediction
- classification
- metrics
- machine learning

REFERENCES

1. Handa, A.; Sharma, A.; Shukla, S. Machine Learning in Cybersecurity: A Review. *Wiley Interdiscip. Rev. Data Min. Knowl. Discov.* **2019**, *9*, e1306. 10.1002/widm.1306.
2. Deliu, I.; Leichter, C.; Franke, K. In *Collecting Cyber Threat Intelligence from Hacker Forums via a Two-Stage, Hybrid Process using Support Vector Machines and Latent Dirichlet Allocation*, In 2018 IEEE International Conference on Big Data (Big Data), 2018; pp 5008–5013.
3. Roopak, M.; Yun Tian, G.; Chambers, J. In *Deep Learning Models for Cyber Security in IoT Networks*, 2019 IEEE 9th Annual Computing and Communication Workshop and Conference (CCWC), 2019; pp 0452–0457.
4. Abdullah, M. S.; Zainal, A.; Maarof, M. A.; NizamKassim, M. In *Cyber-Attack Features for Detecting Cyber Threat Incidents from Online News*, 2018 Cyber Resilience Conference (CRC), 2018; pp 1–4.
5. Sathya, K.; Premalatha, J.; Suwathika, S. In *Reinforcing Cyber World Security with Deep Learning Approaches*, 2020 International Conference on Communication and Signal Processing (ICCSP), 2020; pp 0766–0769.
6. Azwar, H.; Murtaz, M.; Siddique, M.; Rehman, S. In *Intrusion Detection in Secure Network for Cybersecurity Systems using Machine Learning and Data Mining*, 2018 IEEE 5th International Conference on Engineering Technologies and Applied Sciences (ICETAS), 2018; pp 1–9.
7. Kadoguchi, M.; Hayashi, S.; Hashimoto, M.; Otsuka, A. In *Exploring the Dark Web for Cyber Threat Intelligence using Machine Leaning*, 2019 IEEE International Conference on Intelligence and Security Informatics (ISI), 2019; pp 200–202.

8.   Ibrahim, G.; et al. Cybersecurity and Cyber Forensics: Machine Learning Approach. *Mach. Learn. Res.* **2020**, *5* (4), 46–50.

9.   D'hooge, L.; et al. In *In-depth comparative evaluation of supervised machine learning approaches for detection of cybersecurity threats*, 4th International Conference on Internet of Things, Big Data and Security (IoTBDS), 2019.

10. Sebastian, A.; Baier, Harald. In *A Plea for Utilising Synthetic Data when Performing Machine Learning Based Cyber-security Experiments*, Proceedings of the 2014 Workshop on Artificial Intelligent and Security Workshop, 2014.

11. Md Zahangir, A.; Taha, T. M. In *Network Intrusion Detection for Cyber Security using Unsupervised Deep Learning Approaches*, 2017 IEEE National Aerospace and Electronics Conference (NAECON); IEEE, 2017.

12. Giovanni, A.; et al. In *On the Effectiveness of Machine and Deep Learning for Cyber Security*, 2018 10th International Conference on Cyber Conflict (CyCon); IEEE, 2018.

13. Aslan; Çağrı, B.; Sağlam, R. B.; Li, S. In *Automatic Detection of Cyber Security Related Accounts on Online Social Networks: Twitter as an Example*, Proceedings of the 9th International Conference on Social Media and Society, 2018.

14. Boddy, A.; et al. In *A Study into Data Analysis and Visualisation to Increase the Cyber-resilience of Healthcare Infrastructures*, Proceedings of the 1st International Conference on Internet of Things and Machine Learning, 2017.

15. Buczak, A. L.; Guven, E. In *A Survey of Data Mining and Machine Learning Methods for Cyber Security Intrusion Detection*, IEEE Communications Surveys & Tutorials, 2016; vol *18* (2), 1153–1176.

16. Das, R.; Morris, T. H. In *Machine Learning and Cyber Security*, 2017 International Conference on Computer, Electrical & Communication Engineering (ICCECE); IEEE, 2017.

17. Shilp, D.; Kumar, Y. In *Study of Machine and Deep Learning Classifications in Cyber Physical System*, 2020 Third International Conference on Smart Systems and Inventive Technology (ICSSIT); IEEE, 2020.

18. Kamran, S.; et al. Performance Comparison and Current Challenges of using Machine Learning Techniques in Cybersecurity. *Energies* **2020**, *13* (10), 2509.

19. Farooq, H. M.; Otaibi, N. M. In *Optimal Machine Learning Algorithms for Cyber Threat Detection*, 2018 UKSim-AMSS 20th International Conference on Computer Modelling and Simulation (UKSim); IEEE, 2018.

20. Feng, C.; Wu, S.; Liu, N. In *A User-Centric Machine Learning Framework for Cyber Security Operations Center*, 2017 IEEE International Conference on Intelligence and Security Informatics (ISI); IEEE, 2017.

21. Teixeira, M. A.; et al. SCADA System Testbed for Cybersecurity Research using Machine Learning Approach. *Future Int.* **2018**, *10* (8), 76.

22. Torres, J. M.; Comesaña, C. I.; Garcia-Nieto, P. J. Machine Learning Techniques Applied to Cybersecurity. *Int. J. Mach. Learn. Cybern.* **2019**, *10* (10), 2823–2836.

23. Sarker, I. H.; et al. Cybersecurity Data Science: An Overview from Machine Learning Perspective. *J. Big Data* **2020**, *7* (1), 1–29.

24. Sayan, C. M. In *An Intelligent Security Assistant for Cyber Security Operations*, 2017 IEEE 2nd International Workshops on Foundations and Applications of Self* Systems (FAS* W); IEEE, 2017.

25. Fraley, J. B.; Cannady, J. In *The Promise of Machine Learning in Cybersecurity*, SoutheastCon 2017; IEEE, 2017.
26. Shalev, N.; Partush, N. In *Binary Similarity Detection using Machine Learning*, Proceedings of the 13th Workshop on Programming Languages and Analysis for Security, 2018.
27. Sopan, A.; et al. In *Building a Machine Learning Model for the SOC, by the Input from the SOC, and Analyzing it for the SOC*; 2018 IEEE Symposium on Visualization for Cyber Security (VizSec); IEEE, 2018.
28. Yang, X.; et al. Machine Learning and Deep Learning Methods for Cybersecurity. *IEEE Access* **2018**, *6*, 35365–35381.
29. Yavanoglu, O.; Aydos, M. In *A Review on Cyber Security Datasets for Machine Learning Algorithms*, 2017 IEEE International Conference on Big Data (Big Data); IEEE, 2017.
30. Zheng, H.; et al. In *Learning and Applying Ontology for Machine Learning in Cyber Attack Detection*, 2018 17th IEEE International Conference on Trust, Security and Privacy in Computing and Communications/12th IEEE International Conference on Big Data Science and Engineering (TrustCom/BigDataSE); IEEE, 2018.
31. Yamin, M. M.; Ullah, M.; Ullah, H.; Katt, B. Weaponized AI for Cyber Attacks. *J. Inform. Secur. Appl.* **2021**, *57*, 102722.

PART IV
Cybersecurity for Cyber-Physical Systems

CHAPTER 11

EFFICIENT VEHICLE-TRACKING SYSTEM IN DENSE TRAFFIC TO ENHANCE SECURITY

BHARATHI S.,[1] AMISHA R. NAIK,[1] and PIYUSH KUMAR PAREEK[2]

[1]*Department of MCA, Dr Ambedkar Institute of Technology, Bengaluru, India*

[2]*Department of CSE, Nitte Meenakshi Institute of Technology, Bengaluru, India*

ABSTRACT

It is very difficult to identify the vehicle number plate by a surveillance camera when the vehicles are in dense traffic. Because of fast moving of vehicles the image captured by the camera is not visible properly. To reduce this blur Wiener and OCR technique are used. The proposed system gives the robust method for deblurring the blurred number plate images. The objective of the proposed system is that the latent image captured lowers the stress for human eyes in identifying the faded number plate. It is very much required to identify vehicles during accidents or rule-violating vehicles. Because of fast-moving vehicles or during dense traffic snaps taken by a surveillance camera at coverage time are blur. This gives the distorted, unnoticeable, and worsening of photos leading to loss of information, image evidence. In such a situation the proposed algorithm deblurs the image to get useful information from the photos taken for the identification of vehicle's number plate. MATLAB is used to develop the

The Fusion of Artificial Intelligence and Soft Computing Techniques for Cybersecurity.
M. A. Jabbar, Sanju Tiwari, Subhendu Kumar Pani, & Stephen Huang (Eds.)

framework and the output of the work is verified using real-time data. This kind of technique is extensively applied in heavy traffic and high-speed vehicle-moving zones like tolling, no parking zones, etc. The primary objective of this algorithm is designed to improve security system.

11.1   INTRODUCTION

There are various frameworks that are available for automatic number plate recognition in traffic area. This system depends on various procedures for detecting number plate during fast moving of vehicles, uneven erratic number plate, dialect of vehicle number, and variation in lighting circumstances will help to estimate the quality of image. The number plate detection is required to identify the overspeeding vehicle or vehicle violating the traffic rules or during severe accidents. In many situations, the identification of a fast-moving vehicle is detected by using surveillance cameras frequently because of fast-moving vehicles, which is distorted by human eyes. The images captured are blur. There is an extreme loss of edge in low determination.

Due to the fast movement, the number plate image is blurry; nonetheless, the hidden portion is clearly visible thanks to a very often straight uniform convolution and parametrically exhibited length and edge.

Image restoration is a process of set of operations used to achieve deafening an input image to evaluate a noise free and clear image. Clamor encountered in photos are generally salt and pepper noise, Gaussian noise, miss focused photo taken by camera etc. The noises and blur in the image occur due to long time exposure to light which obscures the image. This is the result of comparative movement among camera objects. Automatic number plate recognition framework abnormally detects the vehicle proprietor. It is very useful in detecting problematic vehicles during road accidents or identifying the vehicles which are violating rules, because of the high-speed vehicles, image is blurred due to long exposure time taken by the surveillance camera.

The output is unidentifiable, unnoticeable and debilitation of image gives loss of data. In such circumstances it is better to use image deblurring to get the clue from the image to record car's number plate. The deblurring of such obscured images is done by blind restoration and non-blind restoration. In non-blur restoration the kernel information is well known and in blind image restoration the kernel information is unknown.

The obscured image is mathematically shown as B(a,b) = (K*I)(a,b) + G(a,b) (11.1)

where B represents the blurred image, I is the sharp image, and K is the blur kernel; G is the additive noise (frequently considered as white Gaussian sound) and * represents convolution operator. In blind restoration, kernel K and sharp image I are unidentified.

The blind image restoration problem can be divided into different ways: Uniform BID and nonuniform BID. Nonuniform image restoration is achieved by interchanging the systems which causes the obscurity of image, i.e., first finding the PSF so performing deconvolution whereas in uniform BID finding PSF and image restoration both are done simultaneously.

11.2 METHODOLOGY

The following steps are required to get the required information
1. Acquiring the vehicle image
2. Number plate detection through preprocessing
3. Segmentation of characters
4. Recognition of characters
5. Converting image to text

This framework is implemented and simulated using Matlab R2014a.

The LPR technique is used for image acquisition. In this phase the images are acquired by a purchase system. In this proposed frame, the input image is recorded by a high-resolution camera. The input image size is 1200 × 1600 pixels.

The objective of this proposed system is to enrich the clarity in the image data and to reduce distortions of the image or enrich the image features which is important for future processing. The image preprocessing techniques are used to keep size of the neighborhood pixel that is used for the further calculation of the recent pixel illuminations that are as follows,

- transformations of brightness of the pixel
- geometric transformations
- preprocessing techniques use neighborhood of the processed pixel

- image restoration requires the information about the complete image

11.2.1 CHARACTER SEGMENTATION

Number Plate Segmentation is nothing but character isolation recognizing the region of interest and trying to split it into individual characters. In this work segmentation is carried out through OCR.

11.2.2 CHARACTER RECOGNITION

The mechanisms that allow humans to identify the objects are still under development. The basic principles which are already known by scientists are integrity, purposefulness, and flexibility. The OCR develops the, enabling it to replicate acknowledgments that are typical or human-like. Optical character recognition provides computer vision functions used in an application to perform OCR. OCR is that technique through which the machine vision software reads text and/or characters in the image.

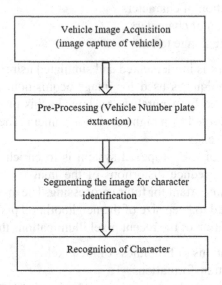

FIGURE 11.1   Methodology.

The process of detecting the characters in the image is commonly known as character segmentation. Before training the characters, the OCR technique gives the standards that segment the characters. Once the segmentation of the characters is over, it stores the knowledge that allows OCR to acknowledge the similar information in other images. The OCR technique is trained by giving the value for segmented characters, generating a singular representation of every segmented character.

11.3 RELATED WORK

The importance of image restoration is explained. It is the group of operations that act on the noisy image to measure the clarity of the image. Noises identified in the image captured are usually salt-and-pepper noise, Gaussian noise, miss focused image captured by camera etc. The image is obscured because of the continuous exposure time. This is the result of relative motion between camera, objects, and scene.[1] Number plate recognition helps in identifying the owner of the vehicle. This is very much useful in recognizing vehicles which are problematic during road accidents or any vehicle violating traffic rules. The image captured during fast-moving vehicle becomes blur. The output image will be undetectable, unrecognizable, and deterioration of image which in turn causes the loss of information. During such situation image deblurring is the best solution to get back the information from the captured image to identify the vehicle's number plate.

Unique registered number plate may be one of indications to identify fast-moving vehicles or hit-and-run case vehicles.[2] The fast-moving vehicle's image captured by surveillance camera is usually blur because of fast movement, which is not even recognizable by human eyes. The images are low intensity and agonies, loss of edge evidence. This requires sightless deblurring techniques. The uniform convolution is required to analyze the blur kernel and modeled parametrically using length and angles. Sparse representation is used to identify blur kernel. During transmission process, image is agonized by several factors, which ends up in some differences with the initial image. The image pre-processing is very important to remove the unwanted information from the captured image and also to improve the image quality by improving the boundary effect and increasing the brightness boundary.[3] After removing noisy part, the image of vehicle's number plate becomes a clear image. Micro-position and height point detection method is used to indulge the position of the kernel. The angle of the kernel is defined by the coefficient of sparse representation of the enhanced image. In the Fourier domain, the kernel's motion's dimension is assessed using random variation.

The methods for deblurring the image are tested blind imaged blurring algorithms shown in this work. This paper is concluded with identifying the trouble maker vehicle by deblurring the obscured image based on the unique ID of the vehicle.[4]

The number plate recognition system will be used for maintaining security issues in parking areas, to control the vehicles in the border trying to cross the unauthorized areas and for identifying the stolen vehicles.[5] These applications will be used in intelligent system to identify the vehicles by using automatic number plate detection in a real-time scenario.

11.4   PROBLEM DEFINITION

Number plate detection is a very significant issue. It is very difficult to identify the owner who is violating the traffic rules. It is required to identify the number plate from the fast-moving vehicle or vehicle in the dense traffic. Automatic Number plate Recognition (ANPR) framework is the better solution for this problem.

The main aim of the proposed system is to decrease crime encountered by vehicles, reduce theft in the shops, decrease the damages, to reinforce the accuracy of the system.

11.4.1   ALGORITHM

TABLE 11.1   Algorithm **for Deblurring the Number Plate.**

```
#Image acquisition(blurred image)
    [fnpn]=uigetfile('*.jpg;','Select image to be blurred');
    I = imcrop(imread([pnfn]));
    I=im2double(I);
imshow(I);
#Preprocessing the blurred image
```

```
PPB = fspecial('motion', 50,45);
Blurred_img = imfilter(I, PSF, 'conv', 'circular');
    figure, imshow(blurred)
imwrite(blurred,strcat('blurre',fn));
figure(5),subplot(1,3,1),
imshow(imagen);
    if size(imagen,3)==3
        imagen=rgb2gray(imagen);
        figure(5),subplot(1,3,2),imshow(imagen),title('Gray Scale Image')
end
threshold_img = graythresh(imagen1);
    imagen1 =~im2bw(imagen,threshold);
    figure(5),subplot(1,3,3),imshow(imagen),title('Black and White
    image')
  imagen = bwareaopen(imagen,30);
#Convert image to text
oc=ocr(imgn);
oc.Text;
#Deblurred image
axes(handles.axes3)
imshow(im2bw(cp,0.8)),title('PreProcessed Image')
bw=im2bw(cp);
st=strel('rectangle',[3 3]);
bwe=imerode(bw,st);
tx=ocr(bwe);
np=tx.Text;
ocr_crp(cp)
```

11.5 RESULT AND PERFORMANCE ANALYSIS

The proposed work involves the following phases:

- Extraction of the region of interest of number plate
- Segmentation of numbers and recognition of numbers

To extract the numbers from the number plate edge detection technique and vertical projection techniques are used. For identifying the numbers diminishing vertical and horizontal estimation, filtering techniques are mainly used. Chain code concept is used to recognize the character with different parameters. The performance of the algorithm is tested with real-time images.

11.5.1 DEBLURRING THE NUMBER PLATE OF MOVING VEHICLES

There are different possible problems in deblurring the image. These include:

1. Image resolution is very poor because the number plate is captured from very far distance or the vehicle is moving very fast, or a low-quality camera is used.
2. Due to poor lighting and low contrast, overexposure, reflection, or shadows.
3. A number on the plate is obscuring due to dirty on the plate.

This proposed framework is developed using MATLAB Software. The proposed framework includes the following steps: Image Acquisition techniques to acquire the picture by a procurement method. In this proposed work, a high-resolution camera is utilized to capture the input image. The captured image considered is 1200 × 1600 pixels. Number Plate Segmentation is mainly used for character isolation which extracts the region of interest and segments the image into individual characters. Number Plate Recognition framework acknowledges the isolated characters. After segmenting the number plate, this method extracts individual characters from the image. Optical Character Recognition technique is used for the recognition of characters from the blurred image.

The initial step is to take out the number plate from the captured image. This is very important to increase the accuracy significantly. The region of interest is extracted using different techniques which include masking of district of number plate with high probability.

11.5.2   DESIGN AND IMPLEMENTATION

Image restoration is done by applying different filters to enhance the noisy and degraded images. The output may not be accurate and varies depending on the limitations of filters. The main objective of the restoration is to remove the degradation which is occurred during image acquisition. Image degradation contains noises which are nothing but errors in pixel values or image is blurred due to the camera motion or disturbances during image acquisition. The image restoration technique enhances the image clarity.

11.5.3 EDGE DETECTION METHODS

Edge detection is very important for object detection in image analysis. Computer vision algorithms are used to implement edge detection. During edge detection process different kinds of breaches occurred in the gray image are nothing but lines, edges, and points line. It is difficult to identify spatial masks with these discontinuities. Edge detection is one of the techniques for detecting important discontinuities in intensity level. The changes in the intensity value of the image are called edges. Edges are generated on the borderline of the two areas. The shape of an object is extracted from the edges of a picture. Edge-identifying techniques are nothing but changes in the original images into boundary images which results in the changes of gray level in the image. The sting identification treats the changes in the gray value of the image and identifies the physical and geometrical properties of objects.

11.5.4 CHARACTERISTICS OF WIENER FILTERS ARE AS FOLLOWS

Wiener filter is used in this proposed work to eliminate the noise in the image. This is a statistical method.

1. Hypothesis: Signal and sound are motionless linear stochastic progressions with spectral features or known autocorrelation and cross-correlation
2. Requirement: The filter must be physically realizable/causal (requirements are frequently dropped, leading to a non-causal solution).
3. Performance criterion: Minimum mean-square error (MMSE).

Wiener filtering technique is based on the deconvolution techniques. When the image quality is degraded by additive noise and blurring, this technique gives the Minimum Square Error. This filtering technique requires signal and noise processes as second-order stationary for calculation. This technique works on the principle of frequency domain. When frequencies have a poor signal-to-noise ratio, this technique tries to attenuate the effect of deconvoluted noise. This technique considers the images and noise as random process.

The technique identifies the noise as random processes, and therefore the purpose is to look out an estimate $\hat{f}$ of the clear image f such that the mean square fault among them is reduced and error measure is specified by

$$e2 = \{(f-f')^2\}$$

where $E\{.\}$ is the arithmetic mean of the argument. The minimum of the error function is given within the frequency domain, by the expression

$$F^\wedge(u,v) = [1H(u,v)\ |H(u,v)|2|H(u,v)|2 + Sn(u,v)Sf(u,v)]G(u,v)$$

where
H(u, v) = degradation function
H*(u,v) = complex conjugate of H(u, v)
(u,v)|2 = H*(u,v) H(u, v)
(u,v) = power spectrum of the noise
(u,v) = power spectrum of the under graded image.

If the noise (u,)/(u,v) is zero then the noise power spectrum vanishes and therefore this method decreases to the inverse filter. If the noise (u,)/(u,v) is large, this filter becomes 0, in order that frequency is ignored. The most advantage is that it takes short computational time to get solution. The main advantage of wiener filter is that it regulates the output error. Sometimes, the resulting outputs are too blurred and spatially invariant.

The LPR image acquisition technique is used for acquiring images. A high-resolution camera is used to acquire the input image 1200 × 1600 pixels, input images are accumulated.

Image representation is a combination of segmented raw pixels. This phase has the methods to show the region supported boundary or group of pixels forming the region. This phase is a kind of classification.

11.5.5   NUMBER PLATE SEGMENTATION

Image segmentation is a very important phase in number plate recognition. The digital images are partitioning into number of regions or group of pixels. The result of image segmentation is contours extracted from number plate image. The image segmentation is done based on similar characters like color, intensity, and texture.

The different image segmentation approaches are,

- Identifying the boundaries between groups which support discontinuities in intensity levels,
- The distribution of pixel properties by using thresholds, like intensity values.
- Directly identifying the region of interest.

Image segmentation depends on the quality of the image. In region-based segmentation techniques, the image is divided into subregions based on the equivalent gray level. Regions are formed based on the common patterns in the intensity values during clustering process.

The easiest segmentation technique is threshold technique. In this technique regions are formed based on the range values. Thresholding is the technique of transforming the input image into a binary image. This technique identifies the regions even if there are any immediate changes in the intensity value. This system is called Edge or Boundary-based techniques; this is one of the most important and fundamental methods in image analysis.

There are techniques used to locate the discontinuities in gray images. The most important approach in edge detection is to identify consequential discontinuities in the gray image. Image segmentation is mainly used to perceive discontinuities in the border. Number plate segmentation is nothing but character segregation, which gets the region of interest and tries to separate each letter. Image segmentation is done through the OCR. Number Plate Extraction may be an important step in the proposed framework, giving the optimized result of the algorithm remarkably. This step takes out the region of interest; that is number plate from the image is captured.

11.5.5.1 NUMBER PLATE RECOGNITION

The LPR technique is applied to acknowledge the separation of characters. After segmenting the extracted numbers into separate character images, the individual characters in an image are recognized. In this proposed work optical character recognition is used for character identification.

11.5.5.2 APPLICATIONS

- To identify vehicle which is very fast over longer distances.
- While crossing the border.
- In petrol stations when a driver goes away without paying for the fuel.
- System for managing the traffic.
- Traveler's behavior analysis for transport planning purposes.
- To identify guest vehicles in visitor management systems
- Auxiliary police use to identify the vehicle which violets the traffic rules.
- Enforcing give way laws for emergency vehicles.

11.5.6 RESULT AND PERFORMANCE ANALYSIS

The input image contains several noises. Noises are removed using techniques such as Weiner filter and OCR to obtain the deblurred image. The output image is in the human readable form. Steps involved in getting the deblurred output image from the blurred input image are as follows:

Step 1: GUI is created which contains the operating buttons within it.

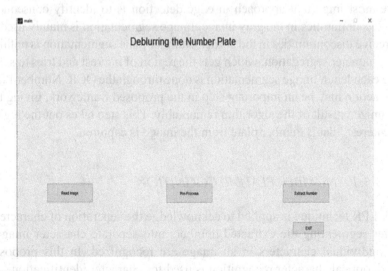

FIGURE 11.2   Creation of GUI.

Step 2: Image is taken on the user interface and the characters are read

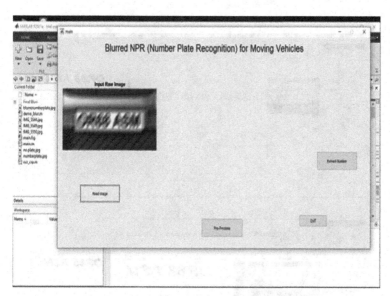

FIGURE 11.3 Reading characters.

Step 3: After reading the image on user interface each character is segmented

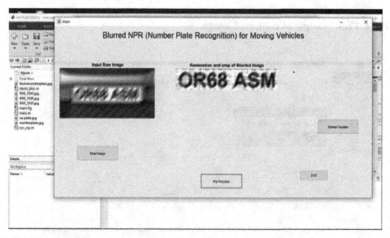

FIGURE 11.4 Character segmentation.

Step 4: In the extraction step each individual character is obtained and displayed on the screen

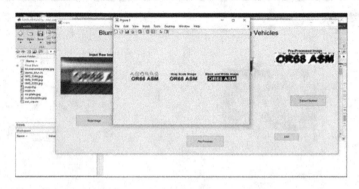

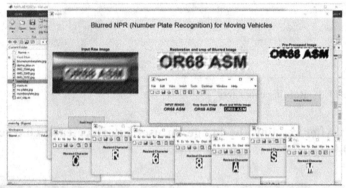

FIGURE 11.5 Output of deblurred image.

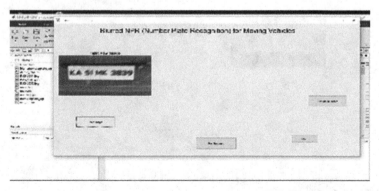

FIGURE 11.6 *(Continued)*

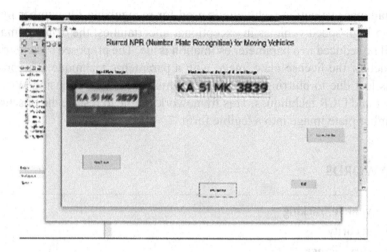

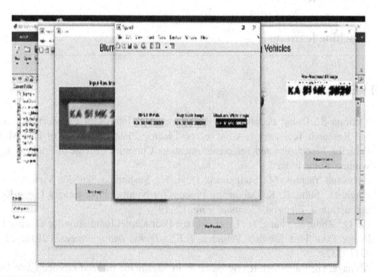

FIGURE 11.6 Output of deblurred image.

11.6 CONCLUSION

Vehicles in the traffic are identified by capturing the vehicle number plate and getting evidence from the vehicle number plate information. The circumstances become complicated when there are a large number of vehicles being identified in various places. In this proposed work, a parameter evaluation technique is used for recognizing for number plate from high-speed vehicles in exceptional uncertainties; the denoise image will be reduced to a parameter estimation issue. The proposed system tried to deblur the license plate image with a parametric technique. The proof gets lost due to blurring of the image which is restored by using Wiener filter and OCR techniques. This framework is able to restore the obscured number plate image into a legible form.

KEYWORDS

- vehicle tracking
- security
- surveillance
- character recognition
- machine learning

REFERENCES

1. Namrata S., Bolaj; Padalkar, G. R. International Journal of Engineering Science and Research Technology. A Survey on License Plate Deblurring of Fast Moving Vehicles. Electronics and Telecommunication, Cummins College of Engineering for Women, Pune, India.
2. Rishwana Yasmin, M.; Sudharsana Devi, K.; Sushmitha Sen, R.; Victoria Dejjee Singh, D.; Sahu, R. K. Deblurring Process for Number Plate Images Using Kernel Estimation. *J. Electron. Commun. Eng.* 15–19
3. Lu, Q.; Zhou, W.; Fang, L.; Li, H. Robust Blur Kernel Estimation for Licence Plate Image from Fast Moving Vehicles. *IEEE. Trans. Image Process.* **2016**, *25* (5), 2311–2323.
4. Abinaya, G.; Banumati, R.; Seshasri, V. In *Rectify the Blurred Licence Plate Image from Fast Moving Vehicles using Morphological Process.* 2017 IEEE International Conference on Electrical, Instrumentation and Communication Engineering (ICEICE), Karur, India, 2017, pp. 1–6
5. Chittode, J. S.; Kate, R. Number Plate Recognition Using Segmentation, *Int. J. Eng. Res. Tech.* **2013**, 1.
6. Chunyu, C.; Fucheg, W.; Baozhi, C.; Chen, Z. In *Application of Image Processing to the Vehicle Number Plate Recognition*, International Conference on Computer Science and Electronics Engineering, Published by Allantis Press, 2013, pp. 2867–2869.
7. Gupta, P.; Purohit, G. N.; Rathore, M. Number Plate Extraction Using Template Matching. *Int. J. Comput. Appl.* **2014**, *88* (3).

CHAPTER 12

OPTIMIZED ANALYSIS OF NETWORK FORENSIC ATTACKS USING AN ENHANCED GROWING NEURAL GAS (GNG) CLUSTERING TECHNIQUE

A. DHANU SASWANTH,[1] V. KAVITHA,[2]
B. SUNDARAVADIVAZHAGAN,[3] and R. KARTHIKEYAN[4]

[1]*Department of Computer Science with Cognitive Systems Sri Ramakrishna College of Arts and Science Coimbatore, India*

[2]*PG & Research Department of Computer Applications, Hindusthan College of Arts and Science, Coimbatore, India*

[3]*Faculty of IT, Department of Information Technology, University of Technology Applied Science-Al Mussanah, Oman*

[4]*Vardhaman College of Engineering, Hyderabad, India*

ABSTRACT

With the enlightenment of the most modern technology intrusion in the domain of cybercrimes as well as networking are escalating at a gradual speed. It leads to boosting the online attacks and crimes in which malicious packets are being sent to further hosts. The concept of Network Traffic Analysis comes under one of the cyber forensics of Network Forensics that handle with recording, capturing, monitoring, and analyzing of network traffic. Network forensics is an extension integrated with the design of

The Fusion of Artificial Intelligence and Soft Computing Techniques for Cybersecurity.
M. A. Jabbar, Sanju Tiwari, Subhendu Kumar Pani, & Stephen Huang (Eds.)

network which classically highlights detection as well as avoidance of society attacks. It will assist the firms to scrutinize the external community. The techniques of Network Forensic can also be utilized to stay away from future attacks. Contemporary techniques of network forensic deal with numerous disputes that must be determined to progress the forensic methodologies. A number of key disputes contain elevated storage speed, data integrity, necessity of plentiful storage space, information privacy, location of information extraction, and access to IP address. Generally, the tools and technologies of network forensic could not be enhanced without addressing the obstacles of the forensic network.

Foremost challenges in network forensics are the ever-growing volume of data that needs to be investigated. This problem has become even more prominent with the emergence of big data and calls for a rethink on the way network forensics investigations have been handled over the earlier period. The machine learning techniques of Growing Neural Gas (GNG) clustering technique have been accepted according to the objective datasets and achievement techniques while achieving forensic investigations. Adding convenience to the network's infrastructure and gathering proof from the intruder using network forensic techniques will help businesses better convey knowledge about network attacks and enable them to investigate both internal and external sources of network security attacks.

12.1   INTRODUCTION

Few years past, most of the crimes had proof that related to the real world. At the present time, the evidence of digital has become of dominant importance. Consequently, forensic sciences expanded their scope to embrace digital evidence. A most important challenge in this field is surviving with gigantic amounts of information during examinations now that the inclination is that the whole thing has become digital procedure. Cyber Forensics is comparatively novel as a discipline of scientific and handles with the acquirement, verification, and investigation of digital evidence.[18] One of the prevalent challenges in this field has thus far been real data sources that are obtainable for experimentation.

Cyber forensics is also referred as digital forensics. "Digital forensics" is the progression of occupying systematic principles along with the procedures to evaluate automatically stored statistics and establish the succession of proceedings which led to a specific occurrence.[19] Through the current digital era, it is very significant for investigators to become conscious of the current enlargements in this energetic domain and recognize scope for the expectations. The ancient decade has eyewitness important scientific improvements to assist during digital exploration. A lot of tactics, tools, as well as techniques have established their way into the domain considered on the standards of forensic. "Digital forensics" has also proved numerous unique approaches that have been included and discovered to attain and investigate digital witness as of miscellaneous sources.[20]

Computer forensics is the domain of examination as well as investigation procedures to assemble and conserve substantiation from a meticulous estimating mechanism in a direction that is appropriate for arrangement in a court of rule. The objective of "Computer forensics" is to achieve a planned exploration and sustain a predictable series of witness to discover accurately what occurred on an estimating machine and who was dependable for it. The expression of "Digital forensics" as well as "Cyber forensics" is frequently utilized as meanings for "Computer forensics."[21]

"Digital forensics" imitate among the gathering of statistics in a direction that preserves its truthfulness. Researchers then scrutinize the facts or scheme to establish if it was transformed, how it was altered and who prepared the modifications. The purpose of "Computer forensics" is not continuously attached to an offense. The forensic procedure is furthermore performed as element of information recovery procedures to collect facts from a damaged server, abortive force, redesigned "operating system (OS)" or further position like a scheme has unpredictably blocked processing.[22]

Depending on universal along with illegal justice scheme, "Cyber forensics" assist makes sure the truthfulness of digital witness offered in civil cases. Digital witness and the forensic process used to gather, safeguard, and analyze it have become more crucial in resolving errors and other legal challenges, much like computers and other information gathering devices are used more frequently in every aspect of life. Digital witness is not now helpful in resolving digital-world crimes, similar to information theft, network breaches, as well as illegal online communications. It is also utilized to explain real-case offense like "burglary, assault, hit-and-run accidents as well as murder."

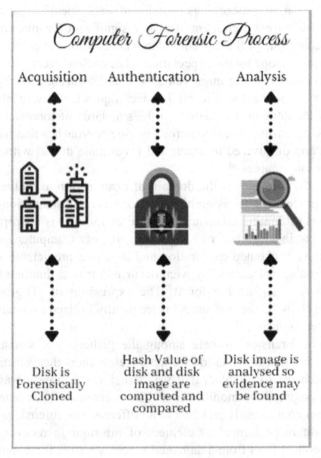

FIGURE 12.1   Conventional computer forensics process.

The conventional "cyber forensics" process is articulated in the procedures of Acquisition, Authentication, as well as analysis as shown in Figure 12.1. Through this traditional procedure of a computer hard device drive is linked to a hardware blocker along with cloned in order to create a bit stream valid duplicate of the hard driver. Further a hash worth value is determined for the disk as well as the copy of forensic and also the hash worth values are evaluated. If the hash worth values match, then the structure is estimated with its genuine and is acknowledged because it has sustained its reliability. Subsequent of this procedure the investigation of the media originates to locate digital witness relatable to the case at hand.

The word "Cyber" termed as "Cybernetics" which denotes the science of perceptive, the control and association of machines as well as animals. Presently, cyberspace is the expression that is almost entirely utilized to portray information security substances. For the reason that, it is very tough to visualize how digital signals roaming across a wire can symbolize an attack to visualize the digital occurrence as a substantial one, cyber threats are a great challenge. Cyberattacks can arise due to crash of military equipment, electrical collapses, and breaches of national security secrets. Cyber security threats occur in three expansive categories of purpose.[19] Cyber threats are never static, it is always dynamic. A cyber security threat is a **cruel** and **purposeful attack** by a human being or association to achieve illegal right of entry to another human being or else concern's network to damage, suspend, or steal Information Technology assets, academic property, computer networks, or any other type of insightful statistics.

12.1.1 VARIETIES OF "CYBER SECURITY THREATS"

At the same time as the classifications of "Cyber Threats" persist to cultivate, there are few of the major frequent as well as widespread cyber threats that present business scenario requires to identify.[16] Some of the cyberattacks are as follows:

➤ **"Malware"**

"Malware attacks" are the very frequent and universal kind of cyberattack, which is referred as "malicious software, together with spyware, ransomware and viruses along with worms," which acquires established into the scheme when the user connects a hazardous connect or mail. Formerly within the system, malware can block rights to dangerous modules of the network, smash up the scheme, and assemble private statistics, surrounded by others.

➤ **"Phishing"**

Cyber criminals send nasty emails that seem to come from genuine along with justifiable properties. The consumer is then scammed into connecting the cruel connection in the electronic mail, leading to malware mechanism or discovery of susceptible data such as credit card particulars information along with login identification.

> **"Spear Phishing"**

"Spear phishing" is a further complicated structure of a phishing attack in which cyber criminals' intention is simply privileged consumers like system administrators as well as "C-suite" executives.

> **"Man in the Middle Attack"**

"Man in the Middle (MitM) attack" happens when cyber illegal's position themselves among multiple gathering communications. Formerly, the attackers understand the statement, they might strain an appropriate susceptible information along with revisit unusual reply to the clients.

> **"Denial of Service Attack"**

"Denial of Service attacks" aspires to flooding procedures, networks, or servers with enormous congestion, in that way creating the system not capable to accomplish lawful demands. Attacks can furthermore employ numerous damaged mechanisms to initiate an attack on the objective structure. This is recognized as a "Distributed Denial of Service (DDoS) attack."

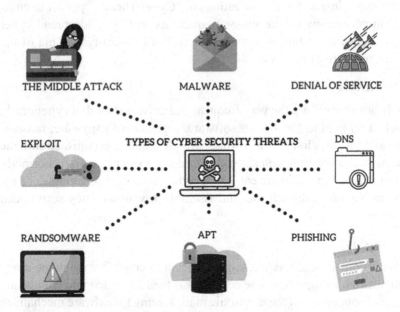

FIGURE 12.2   Types of cyber security threats.

> **"SQL Injection"**

The "Structured Query Language (SQL) injection attack" happens as soon as cyber criminals challenge to access the database through uploading the kind of hateful scripts of SQL. Formerly prosperous, the malevolent performer can modify, outlook, or remove information stored in the SQL database table.

> **"Zero-day Exploit"**

The "Zero-day Exploit attack" happens while either "software vulner-ability" or "hardware vulnerability" is proclaimed, and furthermore the cyber criminals develop the helplessness before a bit or explanation is enhanced.

> **"Advanced Persistent Threats (APT)"**

The "Advanced Persistent Threat" ensues while a hateful performer expands unconstitutional access to a scheme or network and remains unobserved for an unlimited schedule.

> **"Ransomware"**

"Ransomware" is one of the kinds of malware attack in which the aggressor encrypts or locks the attacker's information as well as pressures to distribute or block access to statistics except a ransom is remunerated.

> **"DNS Attack"**

The "DNS attack" is another kind of cyberattack in which cyber crimi-nals develop susceptibilities in the "Domain Name system (DNS)." The attackers influence the DNS vulnerabilities to deflect site guests to spiteful page like "DNS Hijacking" in addition to exhilarating statistics from compromised procedures.

12.1.1.1 NETWORK FORENSICS

With 3800 testimony data breaches in the year 2019 where 89% of them were the attackers from outside concerns should be raised on developing computes to keep away from cyberattacks and loss of meaningful infor-mation. Setting up a firm's network security procedure is not as straight-forward as establishing an antivirus resolution in the individual computer; there is more to it than that.

Network forensics is a scientific technique that centers on the detection and retrieval of data surrounding a cybercrime within a networked atmosphere. General forensic activities comprise the capture, recording, and analysis of occurrences that happened on a network in order to ascertain the basis of cyberattacks. "Network forensics" is the progression of observing packs as well as analyzing network traffic movement for interruption or malware detection. It absorbs categorizing an issue, collecting and investigating data, deciding on the finest troubleshooting reply, and developing it.

"Network forensics or tools of network forensics" are classically utilized into multiple schemes to achieve information gathering and investigation: The "catch it as you can" scheme, anywhere all the information passing through the network is gathered and observed, and the "stop, look, and listen" scheme, where all information container is observed and simply the apprehensive information is detained and evaluated further.[17] While efficient, the primary scheme employs a considerable size of storage; the secondary scheme, on the other hand, does not need as great storage space, but it needs a quicker along with further authoritative processor.

Most of the disputes embrace elevated storage speed, the necessity of sufficient storage gap, data reliability, data confidentiality, access to IP address, and position of information extraction. The features regarding these confronts are afforded with probable explanations to these disputes.

12.1.1.2   GROWING NEURAL GAS CLUSTERING TECHNIQUE

The "Growing Neural Gas" approach involves changing the amount of neurons at a specific location, which is not an unchallengeable input limitation but is changed during the struggle. The correlations among neurons are also not everlasting.[2] As a result of the competition, a set of distinct neural networks covering a specific portion of the input data could be created.

Growing Neural Gas is a simple technique that, like other competing models, allows for topology learning and representation. It is thus a type of topology-representing network, capable of approximating something as complicated as the topology of a human face while drastically lowering information and creating something relatively recognizable to a human observer. For example, the animation to the left shows how complicated something as intricate as the human face can be.

The "Growing neural gas (GNG)" is the approach of "unsupervised topology learning" that copies an information space using interrelated units that are located in the majority densely occupied locations. It produces a graph that may be shown on a 2D plane and reveals cluster prototypes in datasets. When qualified on high-dimensional information, however, GNG frequently produces densely connected graphs, resulting in excessively congested 2D illustrations that might fail to reveal relevant outlines. Furthermore, its chronological learning limits its ability to execute quicker on local datasets and, more crucially, to train on dispersed datasets using the computational capabilities of the infrastructures in which they reside.[7]

Data sets are becoming increasingly large and complicated. Clustering is utilized to expose formations as well as to discover "natural" clustering in multivariate information, and is regarded the most significant unsupervised learning task. In data mining applications, data clustering methods are commonly employed. The goal is to sort N unlabeled items into groups based on their commonalities in a P-dimensional space. Traditional methods such as k-means, on the other hand, enforce a formation happening the information rather than discovering it. The primary scheme is to arrange prototypes into clusters with members that are meaningfully comparable.

The Growing Neural Gas algorithm forms network topology exclusive of being constrained by k Dimensional structures by using the GCS growth process linked with Competitive Hebbian Learning (CHL). The neighborhood relationship between neurons is generated by connecting them, which happens among the first as well as second winning neurons in every cycle. Novel neurons are established in the appropriate order, each with the gathered error index, in the hopes of reducing errors in subsequent iterations.[5] This GNG guidance dynamics generates a structure in the production space that can be thought of as a chart, including neurons acting as vertices. CHL creates associations between surrounding units, which serve as the graph's edges. If the information dimension exceeds three, a dimensionality reduction procedure is required.

The "Self-Organizing Map" is a renowned unsupervised neural learning technique (SOM). It has shown to be an effective tool for visualizing high-dimensional data. While neurons aim to estimate the density of incoming information, it conducts a nonlinear projection from high-dimensional information onto a usual (typically) 2D grid. The goal is to create model vectors that can symbolize the input information set while achieving a permanent planning from the input space to a lattice at the same time.

SOM neural networks have been constructed in a variety of ways. In many circumstances, the structure is defined a priori; therefore, approaches should either employ a fixed amount of collections or attempt to determine with the help of "posteriori algorithms." The authors present information clustering algorithm based on "Growing Neural Gas" that is both efficient and effective (GNG). The parallel implementation of GNG resolves the computational difficulty. To efficiently train such a "Neural network using a High Performance Computing (HPC)" cluster using MPI, some technical issues must be handled. The paper also considers the usual serial approach to GNG training. The parallel version of GNG is benchmarked using the serial learning GNG technique.

The diagram "Neural Gas Production Competitive Hebbian Learning" generates and updates a Parallel Approach in real time. New neurons are automatically added based on the pre-set circumstances, and associations among neurons are matter to the constraints of time and can be erased. GNG can be utilized for vector quantization, biologically influenced image compression, and disease detection by locating code-vectors in clusters.

Algorithm for "Growing Neural Gas" Clustering algorithm

STEP 1: Establish the network process with two neurons using weight vectors.

STEP 2: Opt the arbitrary unutilized input data vector.

STEP 3: Achieve the one learning iteration.

STEP 4: Minimize the error value for all neurons

STEP 5: Redo step 2 until the entire input data vectors are planned to use.

STEP 6: Iterate until a few boundary conditions like greatest number of iterations are discovered.

12.2   RESULTS AND DISCUSSIONS

The "UNSW-NB 15 dataset's raw network packets" can be created using the IXIA Perfect Storm program in "UNSW Canberra's Cyber

Range Lab" to create a hybrid of real recent regular tricks and synthetic modern attack behavior. 100 GB of raw traffic was arrested utilizing the tcpdump program (e.g., Pcap files). "Fuzzers, Analysis, Backdoors, DoS, Exploits, Generic, Reconnaissance, Shellcode, and Worms" are surrounded by the 9 categories of attacks in this dataset. To construct a total of 49 characteristics along with the class label, the Argus and Bro-IDS tools are utilized, and 12 methods are built. The network forensic dataset includes information like Duration, Prototype, State, SBytes, Rate, and Sttl. For the GNG clustering implementation, ten datasets are explored.

The network forensic dataset has the following details such as Duration, Prototype, State, SBytes, Rate along with Sttl. Ten datasets are considered for GNG clustering implementation.

TABLE 12.1   Data Set Details of Network Forensics.

Duration	Prototype	State	SBytes	Rate	Sttl
0.000011	udp	INT	496	90,909	254
0.000008	udp	INT	1762	125,000	254
0.000005	udp	INT	1068	200,000	254
0.000006	udp	INT	900	166,667	254
0.00001	udp	INT	2126	100,000	254
0.000003	udp	INT	784	333,333	254
0.000006	udp	INT	1960	166,667	254
0.000028	udp	INT	1384	35,714	254
0	Arp	INT	46	0	0

Table 12.1 explains the details of the data entity like duration of the attack, prototype of the attack, from which state the attack has been made, Sbyte of the attack, rate of the attack along with sttl values.

The above-mentioned figures are explained about the grouping results with the assistance of Growing Neural Gas clustering technique along with the attributes such as Sttl, Prototype, duration, and State.

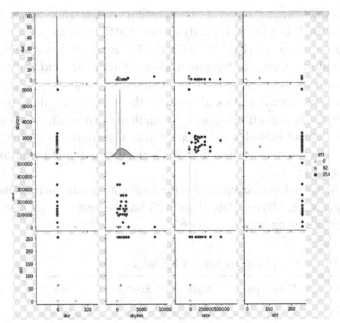

FIGURE 12.3   Cluster the data using the attribute sttl.

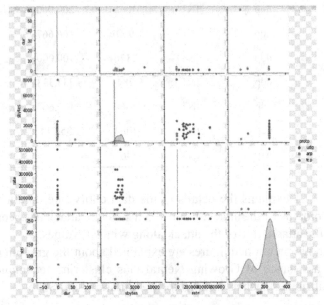

FIGURE 12.4   Cluster the data using the attribute prototype.

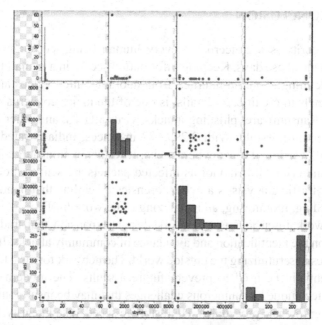

FIGURE 12.5 Cluster the data using the attribute duration.

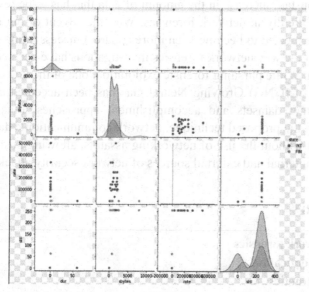

FIGURE 12.6 Cluster the data using the attribute state.

12.3   CONCLUSION

Cyber security is a concern for every human being, corporations, and governments these days. Keeping information secure in a humanity where the whole thing is on the internet, from adorable kitten videos and vacation journals to credit card details, is one of the major concerns of cyber security. Ransomware, phishing attacks, virus attacks, and other types of threats are all possible. With 2,299,682 instances, India is rated 11th in the world in terms of local cyberattacks. It results in an upsurge in online attacks and criminal behavior as infected packets are sent to more hosts. Network traffic analysis is a cyber forensics discipline that handles with the recording, monitoring, and analyzing of network traffic.

"Network forensics" is a network design expansion that traditionally focuses on the identification and avoidance of community attacks. It will aid businesses in scrutinizing the outside world. The network forensic technique can potentially be used to prevent future assaults. The modern network forensic technique has numerous challenges that may be insurmountable in generating forensic procedures. High storage speed, data integrity, the need for a large amount of storage space, data privacy, the location of data extraction, and access to IP addresses are only a few of the major challenges.

The constant increase in the amount of data that has to be reviewed is a major difficulty in network forensics. With the advent of big data, this barrier is expected to become even more apparent, necessitating a reorganization of the way network forensics investigations have been conducted in the past. While pursuing forensic exploration, one of the key clustering strategies of (GNG) Growing Neural Gas has been accepted according to the target datasets and accomplishment approaches. Finally, after employing the proposed technique, network communication data is better connected without the use of networking assaults, allowing businesses to study the internal and external sources of network security attacks.

KEYWORDS

- network forensics
- authentication
- cybernetics
- growing neural gas clustering

REFERENCES

1. Jose Alfredo, F.; Ricardo, C.; Oliceira, S. In *Cluster Analysis using Growing Neural Gas and Graph Partioning*, IEEE 2007 International Joint Conference on Neural Networks, 10.1109/IJCNN.2007.4371447 (accessed Oct, 2007).

2. Šimon, M.; Huraj, L.; Pospíchal, J. Neural Gas Clustering Adapted for Given Size of Clusters doi.org/10.1155/2016/9324793 (accessed Nov, 2016.)

3. Rizzo, R.; Urso, A. In *Identifying Clusters Using Growing Neural Gas: First Results*. Springer International Conference on Artificial Neural Networks ICANN 2009: Artificial Neural Networks – ICANN 2009; pp 536–545

4. Canales, F.; Chacon, M. In *Modification of the Growing Neural Gas Algorithm for Cluster Analysis*. Springer Iberoamerican Congress on Pattern Recognition CIARP 2007: Progress in Pattern Recognition, Image Analysis and Applications, pp 684–693

5. Ultsch. Self-Organizing Neural Networks for Visualization and Classification. In *Information and Classification*, Opitz, O., et al., Eds.; Springer: Berlin, 1993; pp 301–306.

6. Doherty, K. A. J.; Adams, R. G.; Davey, N. Hierarchical Growing Neural Gas. In *Adaptive and Natural Computing Algorithms*; Ribeiro, B., Albrecht, R. F., Dobnikar, A., Pearson, D. W., Steele, N. C. Eds.; Springer: Vienna. https://doi.org/10.1007/3-211-27389-1_34.

7. Vesanto, J.; Alhoniemi, E. Clustering of the Self-Organizing Map. *IEEE Trans. on Neural Networks*, **2000**, *11* (3), 586–602.

8. Silva, M. A. S.; Monteiro, A. M. V.; Medeiros, J. S. In *Semi-Automatic Geospatial Data Clustering by Self-Organizing Maps*. Proc. of the Brazilian Symposium on Neural Networks. Sao Luiz, MA, October, 2004.

9. Fritzke, B. Growing Cell Structures – A Self-organizing Network for Unsupervised and Supervised Learning. *Neural. Netw.* **1994**, *7* (9), 1441–1460.

10. Meghanathan, N.; Allam, S. R.; Moore, L. A. Tools and Techniques for Network Forensics. *Int. J. Netw. Secur. App.* **2009**, *1* (1).

11. Kessler, G. In *Online Education in Computer and Digital Forensics*, Proceedings of the 40th Hawaii International Conference on System Sciences, 2007.

12. Sisaat, K.; Miyamoto, D. In *Source Address Validation Support for Network Forensics*. Proceedings of the 1st Joint Workshop on Information Security, Sep 2006.

13. Mitropoulos, S.; Pastos, D.; Douligers, C. In *Network Forensics: Towards a Classification of Traceback Mechanisms*. Proceedings of the Workshop on Security and Privacy for Emerging Areas in Communication Networks, September 2005, pp 9–16.

14. Ansari, S.; Rajeev, S.; Chandrasekhar, H. Packet Sniffing: A Brief Introduction. *IEEE Potentials* **Dec 2002/Jan 2003**, *21* (5), 17–19.

15. Kavitha, V.; Hemashree, P.; Dilip, H.; Elakkiyarasi, K. Clustering Process with Time Series Data Stream. *Smart Innov. Syst. Tech.* **2020**, *159*, 335–343.

16. Data for Cybersecurity Research: Process and "Whish List". http://www.gtisc.gatech.edu/files_nsf10/data-wishlist.pdf (accessed July 15, 2013).

17. Mahmood, T.; Afzal, U. In *Security Analytics: Big Data Analytics for Cyber Security: A Review of Trends, Techniques and Tools*, Proceedings of the 2nd National

Conference on Information Assurance, Rawalpindi, Pakistan: IEEE, 2013, pp 129–134. doi:10.1109/NCIA.2013.6725337

18. Alvaro, A. C.; Pratyusa, K. M.; Sreeranga, P. R. Big Data Analytics for Security. *IEEE Security and Privacy* **2013**, *11* (6), 74–76. doi:10.1109/MSP.2013.138

19. Berman, J. J. *Principles of Big Data: Preparing, Sharing, and Analyzing Complex Information.* San Francisco, CA: Morgan Kaufmann Publishers Inc, 2013.

20. Sharif, A.; Cooney, S.; Gong, S.; Vitek, D. In *Current Security Threats and Prevention Measures Relating to Cloud Services, Hadoop Concurrent Processing, and Big Data.* 2015 IEEE International Conference on Big Data (Big Data), IEEE, 2015, pp 1865–1870.

21. Rawat, D.; Ghafoor, K. Z. *Smart Cities Cyber Security and Privacy.* Elsevier, December 2018.

22. Constantine, C. Big Data: An Information Security Context. *Netw. Secur.* **2014,** *2014* (1), 18–19. doi:10.1016/S1353-4858(14)70010-8

CHAPTER 13

SECURITY IN IOT: SYSTEMATIC REVIEW

K. TEJASVI, RUQQAIYA BEGUM, and M. A. JABBAR

Vardhman College of Engineering, Hyderabad, India

ABSTRACT

IoT platforms have grown into a worldwide powerhouse over the past ten years that dominates boosting human existence with its uncontrolled smart services throughout every aspect of our daily lives. The ease of accessibility and the rapidly growing need for smart devices and networks are causing IoT to face more security challenges than ever before. By using already-existing security measures, IoT may be safeguarded. The explosions in technology as well as the different attack patterns and their severity make traditional methods less effective. Therefore, the next-generation IoT system needs a robust, constantly improved, and current security system. A considerable technical advancement in machine learning (ML) has created a number of new research opportunities for tackling existing and upcoming IoT challenges. To do this, ML is being utilized as a strong tool to spot assaults and strange behaviors in networks and smart devices. The architecture of IoT is explored in this book chapter after a thorough literature ML methods research that emphasizes the need of IoT security in light of a variety of possible attacks and potential IoT security solutions based on ML.

The Fusion of Artificial Intelligence and Soft Computing Techniques for Cybersecurity.
M. A. Jabbar, Sanju Tiwari, Subhendu Kumar Pani, & Stephen Huang (Eds.)

13.1 INTRODUCTION

An interconnected network of distributed embedded systems called the Internet of Things (IoT) uses wired or wireless communication methods to exchange information.[1] It may alternatively be described as a network of physical objects or objects with network connectivity, software, electronics (such as sensor devices), and limited processing, storage, and communication capabilities that enable them to acquire, sporadically process, and communicate data. In the Internet of Things (IoT), the term "things " refers to everyday objects like smart bulbs, smart adapters, smart meters, smart ovens, AC, IP cameras, and more sophisticated technology like radio frequency identification (RFID) devices, heartbeat detectors, accelerometers, sensors in parking IoT, and various other sensors in cars, etc. The Internet of Things (IoT) provides a wide range of services and applications, including those for military, household appliances, critical infrastructure, agricultural, and personal health care.[3]

IoT services are available in many different industries, such as manufacturing, energy, retail, transportation, and building management. The enormous size of IoT networks creates new difficulties with managing devices, handling vast volumes of data, storing it, communicating it, computing it, and protecting privacy. Research on these multiple IoT characteristics, including as architecture, connectivity, protocols, applications, security, and privacy, has been considerable.[1-4] The assurance of security and privacy as well as customer satisfaction, however, is the foundation of the monetization of IoT technology.[1,4] Because IoT relies on supporting technologies like software-defined networking (SDN), cloud computing (CC), and fog computing, the danger landscape for attackers is expanded.[3]

IoT devices generate a lot of data; thus, traditional techniques for gathering, preserving, and interpreting this data may not be effective at this extent. The sheer amount of data can also be utilized to analyze, forecast, and detect patterns and behaviors. The diversity of data generated by IoT also creates a new application for already-used data processing methods. As a result, new systems are needed to take use of the value of data provided by the IoT. Machine learning (ML) is employed in this context and is thought to be one of the finest computational paradigms for providing integrated cognition in IoT devices.[2] Robots and intelligent machines can benefit from machine learning to help them extrapolate knowledge from data provided by other machines or people. The ability of a smart device to automate or modify a condition or behavior based on information is a crucial component of an IoT solution. For classification, validation, and cluster analysis tasks, machine learning methods have been applied. Computer vision, fraud protection, informatics, malware scanning, authentication, and voice recognition are just a few of the uses for machine learning (ML). ML may also be used with IoT to provide smart services. However, the study's main focus is on the application of machine learning to provide security and privacy services for IoT networks.

13.2 ARCHITECTURE OF IOT

13.2.1 IOT LAYERS

The Internet of Things architecture, which act as a bridge for numerous hardware applications, is being created to link and extend IoT services to every doorstep. In order to transmit and receive various information/ data, the Internet of Things architecture uses a variety of communication protocols, such as RFID, Wi-Fi, Bluetooth broad and broadband wavelength, IEEE 802.15.4, LPWAN, and ZigBee. To better service their most important clients, big high-tech corporations also have their own IoT platforms, such as Microsoft Azure, Amazon AWS IoT, Samsung Artik Cloud, Google Cloud, and others.[1] The three tiers that make up the basic IoT architecture are the network layer, perception/physical layer, and web/ application layer.

13.2.1.1 APPLICATION LAYER

In IoT systems, the third tier is the application layer, which provides services to users via mobile and web-based apps. IoT serves a range of objectives in today's technologically evolved society based on recent developments and applications for smart things. The Internet of Things system and its numerous services have enabled advancements in the smartness of living spaces, transportation, medicine, education, farming, industry, and other sectors.[2,4]

13.2.1.2   NETWORK LAYER

The network layer is very important in Internet of Things (IoT) systems because it connects devices with smart services utilizing a variety of connection protocols.[2] Local clouds and servers store and process data at the network layer, which serves as a bridge between the network and the next levels.[1] Another essential element of the network layer is big data, which has caught the attention of today's rapidly expanding economic sector. IoT systems transmit, analyze, and store a substantial volume of information/data produced by tangible objects at the physical layer. Machine learning (ML) and deep learning (DL) are frequently used to assess information/data that has been saved and generate useful applications for smart devices since information/data is necessary for smart services at the network layer.[2]

13.2.1.3   PERCEPTION LAYER

The perception layer, the first layer of the Internet of Things (IoT) architecture, is composed of the physical (PHY) and media access control (MAC) layers. The PHY layer focuses on the hardware, such as sensors and devices, that employs several communication protocols, including Zigbee, RFID, and Bluetooth.[1-5] The MAC layer creates a connection between networks and physical devices so they can operate together successfully. MAC may connect to network layers using a variety of protocols, including cellular networks. The vast majority of IoT devices are plug-and-play, which generates a lot of huge data.[1-5]

13.3   PROTOCOLS IN IOT

Standards and protocols of IoT are split into two categories.

13.3.1   PROTOCOLS FOR IOT NETWORKS

Devices are connected through the internet using IoT network protocols. These are the communication protocols that are most often used on the Internet. End-to-end data transfer inside the confines of the network is made possible by IoT network protocols. Many IoT network protocols are listed below:

13.3.1.1   HTTP (HYPERTEXT TRANSFER PROTOCOL)

A network protocol for the Internet of Things, such as Hyper Text Transfer Protocol, is the perfect example. It is the foundation of web-based data transfer. When a lot of data has to be released, it is the most popular protocol for Internet of Things devices. On the other hand, the HTTP protocol is not employed because of its high cost, limited energy efficiency, short battery life, and other limitations. Two examples of HTTP protocol application cases are 3D printing and additive manufacturing. It allows computers to connect to a network of 3D printers, allowing for the printing of three-dimensional objects and the processing of prototypes.

13.3.1.2   LONG-RANGE WIDE AREA NETWORK

Long-range, low-power protocol can detect signals despite background noise. Using either private or public networks, battery-powered gadgets may connect to the Internet via the wireless technology known as LoRaWan. A LoRa gateway may be connected to street lighting using this protocol. The gateway then establishes a connection with a cloud application that automates light bulb intensity adjustments based on surrounding lights, reducing power use during the day.

13.3.1.3   BLUETOOTH

Among short-range communication technologies, Bluetooth is one of the most used. This IoT protocol is widely used for wireless data transmission. Short-range, low-cost wireless communication between electronic devices is perfect for this secure communication protocol. Bluetooth is mostly used in mobile devices like smartphones, smart watches, and other gadgets where little amounts of data may be transmitted without using up a lot of power or memory. Of all the IoT device connectivity methods, Bluetooth is the easiest to use.

13.3.1.4   ZIGBEE

Communication between intelligent things is made possible through the Internet of Things (IoT) standard ZigBee. It frequently appears in home automation systems. Electric meters and street lights in urban areas both employ the low-power ZigBee communication technology. Additionally, it is used in security and smart home systems.

13.3.2   DATA PROTOCOLS FOR IOT

IoT data protocols are used to link moderately sized IoT devices. With the use of these protocols, users may communicate with their devices without a network connection. Wired or mobile networks are used to link IoT data protocol connections. The following are a few examples of IoT data protocol:

13.3.2.1   MESSAGE QUEUE TELEMETRY TRANSPORT (MQTT)

MQTT is a well-liked protocol for IoT devices since it collects data from various electrical devices and allows virtual network monitoring. This TCP-based publish protocol enables event-driven message exchange across wireless networks. Most low-cost devices with limited memory and power consumption use MQTT. Among these are text-based apps, smart watches, vehicle sensors, and smoke alarms.

13.3.2.2   CONSTRAINED APPLICATION PROTOCOL (COAP)

A restricted access internet utility protocol is called CoAP. The server may respond to the client's request using HTTP utilizing this protocol, and the client can submit requests to the server. UDP (User Datagram Protocol) is used to implement it quickly and efficiently. The protocol sends a request to the endpoints of the application and then waits for a response from the application's resources and services.

13.3.2.3   ADVANCED MESSAGE QUEUING PROTOCOL (AMQP)

In the context of message-oriented middleware, filtering and processing are accomplished via a software layer protocol known as AMQP. It provides the safe and efficient transfer of data between connected devices and the cloud and is used to set up dependable point-to-point connections. Binding, Exchange, and Message Queue are the three elements of AMQP. The three components work in concert to enable secure communication, transmission, and storage. Additionally, it helps to create the connection between two messages. The majority of banking uses the AMQP protocol. The protocol monitors each message a server sends until it is properly delivered to the target purposes or locations.

13.3.2.4 MACHINE-TO-MACHINE (M2M) COMMUNICATION PROTOCOL

It is an open industry standard for managing remote applications for IoT. It provides a platform for two devices to interact and exchange data. With the help of this protocol, systems may more easily keep an eye on themselves and change to accommodate new conditions. The identification of autonomous vehicles, vending and ATM machines, smart homes, and M2M communication protocols are all employed.

13.3.2.5 EXTENSIBLE MESSAGING AND PRESENCE PROTOCOL (XMPP)

To send and receive messages instantly, XMPP uses a push mechanism. Modifications are simple to implement and XMPP is adaptable. XMPP uses open XML to provide the servers' or devices' availability status while sending or receiving messages in addition to instant chat platforms like Google Talk and WhatsApp. Additionally, XMPP is used by VoIP, gaming, and online news services (VoIP).

13.4 ATTACKS OF IOT

A series of assaults on the IoT system in recent years have made manufacturers and users more careful while designing and deploying IoT devices. We will discuss several attack types, their outcomes, and attack surfaces in the IoT in this section.

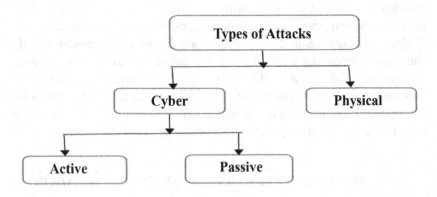

FIGURE 13.1   Various types of IoT attacks.

13.4.1 TYPES OF ATTACKS

The most frequent IoT hazards are physical and cyber assaults, including active and passive cyber-attacks. Various types of attacks are shown in Figure 13.1. Cyber-attacks involve breaking into a wireless network's computer system and changing user data. On the other hand, physical assaults are those that actually damage IoT hardware. In this instance, a network is not necessary for the attackers to attack the system. Because of this, physical IoT devices including cameras, smartphones, routers, sensors, and other devices are vulnerable to these attacks where the attackers disturb the service.[1,2] The next subsections mostly discuss the various cyber-attack types and their severity in relation to IoT security.

13.4.1.1 ACTIVE ATTACKS

When a hacker gains access to the network and the data it holds with the goal of modifying the system's configuration and interfering with certain processes, that is the definition of an active attack. IoT device security may be compromised in a number of ways during active assaults, including intervention, modification, and interruption. Figure 13.1[1] demonstrates current attacks, such as man-in-the-middle, denial-of-service, spoofing, Sybil attack, selective forwarding, jamming, hole attack, data manipulation, and malicious inputs, among others. Active assaults, including denial-of-service, man-in-the-middle, spoofing, jamming, selective forwarding, malicious inputs, and data tampering, etc., are outlined in Table 13.1 and shown in Figure 13.1.[1]

(i) Denial of Service Attacks

The primary reason for DoS attacks is that they repeatedly make duplicate requests, which stops a system from performing its responsibilities. As a result, it is challenging for the user to make an educated decision since they are unable to connect with the IoT device. Additionally, DoS attacks continuously run IoT devices, potentially reducing battery life. A distributed denial of service (DDoS) assault takes place when many attackers use different IP addresses that cause a large number of requests, keeping the server busy. Because of this, it might be difficult to distinguish between legitimate traffic and attack traffic.[1] In recent years, severe DDoS assaults that have destroyed thousands of devices have been brought on by a special IoT botnet virus known as Mirai.[1,2]

(ii) Sybil and Spoofing Attacks

Sybil attacks and spoofing mainly focus on users' identities to gain unauthorized access to the Internet of Things system. IoT devices are particularly vulnerable, particularly to spoofing attacks, since the TCP/IP suite lacks a trustworthy security protocol. These two attacks also sparked more harmful attacks including denial-of-service and man-in-the-middle attacks.[1]

(iii) Jamming Attacks

By transmitting unwanted signals to IoT devices, jamming attacks interfere with current wireless network connectivity and pose problems for users by maintaining a constant level of network activity.[10] By consuming more energy, bandwidth, memory, and other resources, this attack also impairs the performance of IoT devices.

(iv) Man-in-the-Middle Attacks

By using guise of a communication network component, man-in-the-middle attackers assert to be directly linked to another user device. As a result, it may easily stop communications by changing genuine data with fictional and misleading information.[4]

(v) Attacks in Selective Forwarding

In order to create a network hole, the selective forwarding attack functions as a node in the communication system by enabling some data packets to be discarded during transmission. It is challenging to recognize and avoid attacks of this nature.

(vi) Attacks in Malicious Input

Malicious input attacks are malware software assaults like rootkits, and trojans that waste energy, worms, viruses, adware, cause financial loss, and degrade the operation of wireless networks in Internet of Things devices.[1,2]

(vii) Data Tampering

The act of intentionally altering a user's data by an attacker in an effort to violate that user's privacy by engaging in unwanted activities is known as data tampering. Data manipulation attacks are particularly vulnerable to IoT devices that store private user information, such as location, fitness, and payment information for smart equipment.[1]

13.4.1.2   PASSIVE ATTACKS

Passive attacks attempt to collect user information without the individual's consent or agreement, and then use that information to decode their private data.[1] Eavesdropping and traffic analysis are two methods for passively attacking an IoT network. The IoT device of the user is used by eavesdroppers as a sensor to gather and utilize their location and personal information.[1,2]

13.5   INTRUSION DETECTION IN IOT

Due to their diverse nature, considerably bigger typical behavior, and increased vulnerabilities brought on by the exponential proliferation of IoT devices, conventional intrusion detection systems (IDS) are worthless in the IoT environment.[4] As a result, traditional security software cannot be installed in IoT systems due to a lack of computing and storage capacity. Reference[2] claims that safeguarding these devices will need a paradigm shift. Approaches based on signatures and those based on anomalies are the two main intrusion detection techniques used to discover attacks in IoT systems.[1,2] It is occasionally done to combine them to form a hybrid detection system, albeit it is difficult to do.[2] As opposed to anomaly-based systems, which analyze traffic patterns to spot attacks, signature-based solutions categorize traffic as benign or malicious using recognized threats.[2]

On existing and well-known assaults, signature-based solutions perform admirably. One of the drawbacks of the signature-based strategy is the time it takes to keep the signature database up to date. As the database size increases, it becomes computationally expensive to compare input with it. Due to its reliance on previously known attack signatures, this method cannot detect zero-day or undiscovered assaults.[6] When an irregular traffic pattern is identified, an anomaly-based detection technique is utilized because it looks at regular traffic patterns and either raises an alarm or restricts traffic. The ability of anomaly-based systems to detect zero-day and unidentified assaults is one of its benefits, but they may also generate a large number of false positives.[5] One of the main reasons why researchers are concentrating on anomaly-based IDS is their superior ability to identify widespread, zero-day, and unidentified attacks.

13.6 IOT MACHINE LEARNING METHODS

The use of machine learning (ML) as a method of artificial intelligence enables machines to learn by experience rather than via explicit programming.[32] ML can function in dynamic networks and does not require complex mathematical computations or human interaction. Machine learning techniques have made substantial advancements in IoT security during the past several years.[8] Therefore, machine learning approaches could be used to quickly identify various IoT assaults by monitoring device activity. Relevant solutions may also be provided for IoT devices with limited resources by using a variety of ML techniques. To detect clever attacks in IoT devices and develop a robust defense strategy, ML approaches including supervised techniques, unsupervised techniques, and reinforcement learning may be applied.

13.6.1   ML TECHNIQUES

To identify clever attacks on IoT devices and create a powerful security strategy, ML approaches such as supervised techniques, unsupervised techniques, and reinforcement learning may be employed. Figure 13.4 illustrates several machine learning methods for IoT system security.

13.6.1.1   SUPERVISED LEARNING

Supervised learning, which employs a training data set and learning algorithm to classify the output in line with the input, is the most popular kind of machine learning. Both classification and regression learning are supervised learning techniques. Classification Learning: A supervised machine learning technique called classification learning produces categories or fixed discrete values. Examples of several categorization learning strategies, including the Bayesian Theorem, Machine, Association Rule, K-Nearest Neighbor, Support Vector, and Random Forest, will be provided in the subsequent subsections.

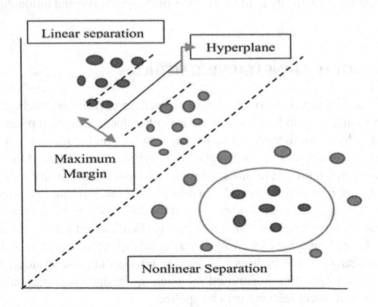

FIGURE 13.2   SVM learning.

(i) Support Vector Machine (SVM)

To analyze data from regression and classification investigations, the SVM technique is employed. A hyperplane, which SVM constructs between two classes, The hyperplane's objective is to maximize distance between each class while minimizing error and maximizing margin (Fig. 13.2).[27-30] After analysis, if the hyperplane is shown to be nonlinear, SVM employs kernel function to make it linear by introducing new features. In SVM, using the optimum kernel function can occasionally be challenging. SVM is useful for IoT security applications such smart grid attacks,[32] intrusion detection,[8,20] malware detection,[31] etc. owing to its high accuracy level.

(ii) The Bayesian Theorem

The base of the Bayesian theorem is known as the Bayesian probability, which is often referred to as the possibility of statistics theorem for learning distribution. By applying Bayesian probability and the provided data, this kind of supervised learning technique generates new outcomes. Nave Bayes is the name of this individual. Because it uses prior theory and Bayesian probability to predict expected events, NB has become popular as a learning algorithm. This is one of the issues that the Internet of Things can effectively resolve. At the network layer, NB is frequently utilized in IoT for anomaly and intrusion detection.[39,40] Some advantages of NB include its simplicity of usage, ability to apply to multi-stage calcification, and need for less categorization details. Because NB depends on characteristics, how those features interact with one another, and previous information, reliable findings might not be possible.[37]

(iii) K-Nearest Neighbor (KNN)

Euclidian distance is frequently used by KNN, a statistically nonparametric supervised learning algorithm.[14] The Euclidian distance in KNN is used to compute the average value of a node's k closest neighbors when it is unknown.[9] For instance, the average value of the closest neighbor may be used to forecast the loss of a node. Even if this number is incorrect, it still helps locate the possible missing node. The KNN method is uncomplicated, reasonably priced, and easy to use.[12,13] On the other hand, locating the missing nodes demands precision that is hard to get and takes a lot of time.

(iv) Random Forest (RF)

A variety of Decision Trees (DTs) are used in RF, a revolutionary machine learning technique, to build an algorithm that produces a reliable and powerful predictive model for outcomes. These several trees are created at random, trained for a particular job, and utilized as the model's final conclusion. Although RF uses DTs, the learning process is different since RF considers the average of the output and needs fewer inputs.[14,15] In network surface attacks, RF is frequently used to identify illegal IoT devices,[18] detect anomalies,[17] and detect DDoD attacks.[16] Previous research demonstrates that RF provides superior outcome in the identification of DDoS attacks using ANN, KNN, and SVM.[16] Although RF cannot be used in real-time applications, more training data sets are needed to create DT's that can recognize unforeseen undesirable intrusions.

(v) Association Rule (AR)

The AR approach, which is used to identify the unknown variable based on how they are connected to one another in a given data set, is another sort of supervised machine learning methodology.[19] To discover the intrusion in the network, fuzzy AR was utilized, which was a successful use of the AR approach for intrusion detection.[40] Despite being equally clear-cut and easy to use, due to its high temporal complexity and reliance on concepts that would not produce a precise result for a massive and complex model, augmented reality (AR) is not commonly employed in the Internet of Things (IoT).[11]

Regression Learning: Corresponding to the input variables, regression learning produces either a real number or a continuous value as the output. In the next subsections, many RLs are given, including Ensemble Learning, Neural Network, and Decision Tree.

(i) Decision Tree (DT)

The branches and leaves of a tree are analogous to the DT technique of natural supervised learning. Various branches act as edges and leaves as nodes in DT. The offered samples are sorted using DTs according to the featured values. Classification and regression are the two primary subcategories of DT in ML.[12] DT is superior to other ML techniques in that it can handle huge data samples, is transparent, and is easy to apply.[13,14] Contrarily, this method has some drawbacks including the need for a considerable amount of storage space because of its substantial structure for the data. If more than one DT is used to solve the issue, the learning procedure becomes more complicated.[13,14] In applications for security like intrusion detection, DTs and DDoS are frequently employed as classifiers.[17,25]

(ii) Neural Network (NN)

The neuron-using structure of the human brain served as the basis for the development of the NN method. ML methods that can handle difficult complex situations are frequently employed in NN.[18,19] Hierarchical and connected networks are the two primary network types used in NN techniques. As shown in Figure 13.3, these two network types are based on the input, hidden, and output layers of the neuron, which have distinct functional levels. The IoT system performs better because NN techniques reduce network response times. Because of their computational complexity, NN are difficult to deploy in a distributed IoT system.

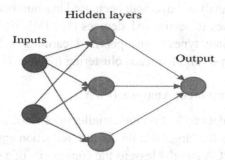

FIGURE 13.3   Neural network.

(iii) Ensemble Learning (EL)

In Machine Learning, a learning algorithm called EL employs a range of classification strategies to enhance its performance and generate an acceptable outcome view (Figure 13.4). To deliver a reliable outcome, EL frequently combines homogeneous or heterogeneous multi-classifier. EL is well suited to address the majority of issues because it makes use of many learning techniques. In contrast to other single-classifier methods, EL has a high temporal complexity. El is frequently utilized for intrusion, malware, and anomaly detection.[20–22]

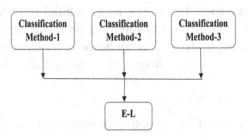

FIGURE 13.4 Ensemble learning.

13.6.1.2 UNSUPERVISED LEARNING

Unsupervised learning occurs when the input variables are not given any output information. While the algorithm searches for patterns in this data set, the majority of the data are unlabeled. On the basis of that, it divides them into various groupings. Infinite Gaussian mixture models and continuous correlation analysis have both been used in a number of unsupervised learning techniques to secure IoT devices (IGMM).[23,24] The next section will go through many types of unsupervised learning, such as the Principal Component Analysis and K-means clustering methods (PCA).

(i) Principal Component Analysis (PCA)

To divide a huge data collection into smaller ones while retaining the same amount of data as the larger set, the feature reduction approach known as PCA is applied. PCA thereby lessens the complexity of a system. A feature to detect real-time intrusions into an IoT system may be selected using this technique.[25] A robust security protocol can be created by combining PCA with a few other ML techniques. A model presented by Ref. [26] creates an effective system using PCA and classifier techniques like KNN and SoftMax regression.

(ii) K-mean Clustering

Using an unsupervised learning approach, the given data samples are divided into smaller groups and then classified as a cluster. This well-known method makes use of clustering strategies. You may use this technique by adhering to a few simple rules: (1) The given data set is initially divided into a number of clusters, with the primary goal being to determine the k-centroid for each cluster. (2) Once all nodes have been reached, choose a node from each cluster and link it to the nearest centroid again. Afterward, a fresh computation is performed using the average value of each node from each cluster; and (3) the algorithm repeats its previous steps until they coincide with the K-mean value.[27–29] In order to find suitable housing, smart cities can benefit greatly from K-mean learning approaches. Because of their simplicity, K-mean algorithms are excellent for IoT systems that do not require tagged data. But the performance of this unsupervised learning system is poorer than supervised learning. The detection of Sybil attacks and anomaly identification are two common applications of the K-mean clustering technique.[30,32–34]

13.6.1.3 *REINFORCEMENT LEARNING (RL)*

By taking actions to optimize the total feedback, through interactions with its surroundings, a computer using RL can learn much like a person.[35,36] Depending on the results of the assigned task, the feedback might be a reward. In reinforcement learning, there are no predetermined behaviors for each particular task; instead, the computer uses methods of iterations. Iterations help the agent choose the optimum strategy to maximize reward while learning from experience. Reinforcement learning is used by many IoT devices (such as sensors, electric windows, and air conditioners) to adapt to their surroundings. In order to detect different IoT assaults and offer the devices with appropriate security protocols, RL approaches have been applied for IoT device security. While Dyna-Q has been employed for malware detection and authentication, Q-learning has been used for authentication, jamming attacks, and harmful inputs.[36–38,40] Additionally, DQN and PDS, respectively, can offer security against jammer assaults and malware detection.[39] IoT Security Using Machine Learning is shown in Table 13.1.

TABLE 13.1   Applications of Machine Learning in Security.

ML method	Application/Attack detection	Ref.
NN	IoT Security	[6]
	DoS	[7]
	Intrusion Detection	[8, 9]
	IoT Privacy	[10]
	Mobile Networks Security	[11]

TABLE 13.1   *(Continued)*

ML method	Application/Attack detection	Ref.
KNN	Intrusion Detection	[12, 13]
	Anomaly, Impersonation Attacks, False Data Injection Attacks, Detection of Intrusion	[92, 15]
	An IoT element's authentication	[16]
SVM	Intrusion Detection	[9, 16–21]
	Mobile Networks Security	[11]
	Authentication, False Data Injection Attacks, Data Tampering, and Abnormal Behavior	[22–24, 92]
DT	Unusual Traffic Sources and Intrusion Detection	[25]
	Intrusion Detection	[26]
EL	Authentication, False Data Injection Attacks, Detection of Intrusion and Malware, and Data Tampering	[22–24, 92]
K-means	Private Data Anonymization in an IoT System and Data Tampering, Abnormal Behavior, Sybil Detection in Industrial WSNs	[23]
	Intrusion Detection	[27]
	Network attack detection	[28]
NB	Intrusion Detection	[26, 29]
	Anomaly Detection	[30]
	IoT Security	[31]
	Traffic Engineering	[32]
RF	Intrusion Detection	[13, 21, 33]
	Unauthorized IoT Devices, Anomalies and DDoS	[34]
PCA	Intrusion Detection, Real-Time Detection System	[35]
RL	DoS	[37]
	Spoofing	[36]
	Eavesdropping	[38]
	Jamming	[39]
	Malware Detection	[40]
AR	Intrusion Detection	[41]

13.7 CONCLUSION

Due to the enormous data generated by IoT devices attacks on IoT are also expanding. Traditional methods to safeguard this data are not efficient. Machine learning methods are widely used to secure IoT devices from various attacks. This chapter explores various machine learning methods to provide security and privacy services for IoT. In this chapter, protocols in IoT are discussed. Various types of security attacks like active attacks and passive attacks were explained. This survey explores various machine learning methods used in order to safeguard IoT devices. This chapter will help researchers, security practitioners, and network administrators to effectively implement machine learning methods in security.

KEYWORDS

- attack surfaces
- challenges
- architecture
- internet of things
- machine learning
- IoT threats
- security solution

REFERENCES

1. Tahsien, S. M.; Karimipour, H.; Spachos, P. Machine Learning Based Solutions for Security of Internet of Things (IoT): A Survey. *J. Netw. Comput. Appl.* **2020**, *161*, 102630.
2. Ahmad, R.; Alsmadi, I. Machine Learning Approaches to IoT Security: A Systematic Literature Review. *Int. Things* **2021**, *14*, 100365.
3. Canedo, J.; Skjellum, A. *Using Machine Learning to Secure IoT Systems*, https://doi.org/10.1109/PST.2016.7906930, 16824896.
4. Hussain, F.; Hussain, R.; Hassan, S. A.; Hossain, E. Machine Learning in IoT Security: Current Solutions and Future Challenges, *IEEE Commun. Surv. Tutor* **2020**, *22*, Third Quarter.

5. Anthi, E.; Williams, L.; Słowińska, M.; Theodorakopoulos, G.; Burnap, P. A Supervised Intrusion Detection System for Smart Home IoT Devices, *IEEE Int. Things J.* **2019,** *6* (5), 9042–9053. doi:https://doi.org/10.1109/JIOT.2019.2926365.

6. Altaf, A.; Abbas, H.; Iqbal, F.; Derhab, A. Trust Models of Internet of Smart Things: A Survey, Open Issues, and Future Directions. *J. Netw. Comput. Appl.* **2019,** *137,* 93111.

7. Kulkarni R. V.; Venayagamoorthy, G. K. In *Neural Network Based Secure Media Access Control Protocol for Wireless Sensor Networks,* Proceedings of the International Joint Conference Neural Networks, Atlanta, GA, 2009, pp 3437–3444.

8. Buczak A. L.; Guven, E. A Survey of Data Mining and Machine Learning Methods for Cyber Security Intrusion Detection. *IEEE Commun. Surv. Tutor* **2015,** *18* (2), 1153–1176.

9. Sedjelmaci, H.; Senouci, S. M.; Al-Bahri, M. In *A Lightweight Anomaly Detection Technique for Low-resource IoT Devices: A Game-theoretic Methodology,* IEEE International Conference on Communications (ICC). 2016, pp 1–6.

10. Jeong, H. J.; Lee, H. J.; Moon, S. M. In *Work-in-progress: Cloud based Machine Learning for IoT Devices with Better Privacy,* 2017 International Conference on Embedded Software (EMSOFT), Seoul, 2017, pp 1–2.

11. Do, V. T.; Engelstad, P.; Feng, B.; Do, T. V. Strengthening Mobile Network Security Using Machine Learning, In *Mobile Web and Intelligent Information Systems*; Younas, M., Awan, I., Kryvinska, N., Strauss, C., Thanh, D. V., Eds.; Springer International Publishing, Cham, 2016, pp 173–183.

12. Branch, J. W.; Giannella, C.; Szymanski, B.; Wolff, R.; Kargupta, H. In-network Outlier Detection in Wireless Sensor Networks. *Knowl. Inform. Syst.* **2013,** *34* (1), 23–54.

13. Narudin, F. A.; Feizollah, A.; Anuar, N. B.; Gani, A. Evaluation of Machine Learning Classifiers for Mobile Malware Detection. *Soft. Comput.* **2016,** *20* (1), 343–357.

14. Chen, F.; Deng, P.; Wan, J.; Zhang, D.; Vasilakos, A.; Rong, X. Data Mining for the Internet of Things: Literature Review and Challenges. *Int. J. Distrib. Sens. Netw.* **2015,** *11,* 431047.

15. Aminanto, M. E.; Kim, K. In *Improving Detection of WiFi Impersonation by Fully Unsupervised Deep Learning.* Information Security Applications: 18th International Workshop, WISA 2017, 2017.

16. Baldini, G.; Giuliani, R.; Steri, G.; Neisse, R. In *Physical Layer Authentication of Internet of Things Wireless Devices through Permutation and Dispersion Entropy,* 2017 Global Internet of Things Summit (GIoTS), Geneva, 2017, pp 1–6.

17. Bamakan, S. M. H.; Wang, H.; Yingjie, T.; Shi, Y. An Effective Intrusion Detection Framework Based on MCLP/SVM Optimized by Time-varying Chaos Particle Swarm Optimization. *Neurocomputing* **2016,** *199,* 90–102.

18. Kabir, E.; Hu, J.; Wang, H.; Zhuo, G. A Novel Statistical Technique for Intrusion Detection Systems. *Future Gener. Comput. Syst.* **2018,** *79,* 303–318.

19. Wang, H.; Gu, J.; Wang, S. An Effective Intrusion Detection Framework Based on SVM with Feature Augmentation. *Knowl. Based Syst.* **2017,** *130*–139.

20. Zissis, D. In *Intelligent Security on the Edge of the Cloud.* International Conference on Engineering, Technology and Innovation, IEEE, Funchal, 2017, pp 1066–1070.

21. Zarpelƒo, B. B.; Miani, R. S.; Kawakani, C. T.; de Alvarenga, S. C. A Survey of Intrusion Detection in Internet of Things. *J. Net. Com. App.* **2017,** *84,* 2537.

22. Shi, C.; Liu, J.; Liu, H.; Chen, Y. In *Smart User Authentication through Actuation of Daily Activities Leveraging WiFi enabled IoT.* Proceedings of the 18th ACM International Symposium on Mobile Ad Hoc Networking and Computing, ACM, 2017, p. 5.

23. Lee, S.-Y.; Wi, S.-r.; Seo, E.; Jung, J.-K.; Chung, T.-M. In *ProFiOt: Abnormal Behavior Profiling (ABP) of IoT Devices Based on a Machine Learning Approach.* Telecommunication Networks and Applications Conference (ITNAC), 2017 27th International, IEEE, 2017, pp 1–6.

24. Nobakht, M.; Sivaraman, V.; Boreli, R. In *A Host-based Intrusion Detection and Mitigation Framework for Smart Home IoT Using OpenFlow.* 2016 11th International Conference on Availability, Reliability and Security (ARES), IEEE, 2016, pp 147–156.

25. Goeschel, K. In *Reducing False Positives in Intrusion Detection Systems Using Data-mining Techniques Utilizing Support Vector Machines, Decision Trees, and Naive Bayes for Off-line Analysis.* SoutheastCon, 2016, IEEE, 2016, pp 1–6.

26. Stroeh, K.; Mauro Madeira, E. R.; Goldenstein, S. K. An Approach to the Correlation of Security Events Based on Machine Learning Techniques. *J. Int. Serv. Appl.* **2013,** *4* (1), 7.

27. Rathore, H.; Jha, S. In *Bio-inspired Machine Learning Based Wireless Sensor Network Security.* 2013 World Congress on Nature and Biologically Inspired Computing. IEEE, Fargo, ND, 2013, 140–146.

28. IoT Analytics. Why the Internet of Things is Called Internet of Things: Definition, History, Disambiguation. https://iot-analytics.com/internet-of-things-defnition/

29. Usama, M.; Qadir, J.; Raza, A.; Arif, H.; Yau, K. A.; Elkhatib, Y.; Hussain, A.; Al-Fuqaha, A. I. Unsupervised Machine Learning for Networking: Techniques, Applications and Research Challenges. *CoRR* **2017,** arXiv:1709.06599.

30. Mehmood, T.; Rais, H. B. M. In *Machine Learning Algorithms in Context of Intrusion Detection.* 3rd International Conference on Computer and Information Sciences (ICCOINS), IEEE, 2016. https://doi.org/10.1109/iccoins.2016.7783243.

31. Jincy, V. J.; Sundararajan, S. Classification Mechanism for IoT Devices Toward Creating a Security Framework. In *Intelligent Distributed Computing;* Buyya, R., Thampi, S. M. Eds., Springer International Publishing, Cham, 2015, pp 265–277.

32. Hogan, M.; Esposito, F. In: *Stochastic Delay Forecasts for Edge Traffic Engineering via Bayesian Networks.* IEEE International Symposium on Network Computing and Applications. IEEE, Cambridge, MA, 2017, pp 1–4.

33. Farnaaz, N.; Jabbar, M. A. Random Forest Modeling for Network Intrusion Detection System. *Procedia Comput. Sci.* **2016,** *89* (Supplement C), 213–217.

34. Miettinen, M.; Marchal, S.; Hafeez, I.; Asokan, N.; Sadeghi, A.-R.; Tarkoma, S. In *IoT Sentinel: Automated Device Type Identification for Security Enforcement in IoT.* 2017 IEEE 37th International Conference on Distributed Computing Systems (ICDCS), IEEE, 2017, pp. 2177–2184.

35. Deng, L.; Li, D.; Yao, X.; Cox, D.; Wang, H. Mobile Network Intrusion Detection for IoT System Based on Transfer Learning Algorithm. *Cluster Comp.* **2018,** 1–16.

36. Xiao, L.; Li, Y.; Han, G.; Liu, G.; Zhuang, W. PHY-layer Spoofing Detection with Reinforcement Learning in Wireless Networks, *IEEE Trans. Veh. Technol.* **2016**, *65* (12), 10037–10047.

37. Li, Y.; Quevedo, D. E.; Dey, S.; Shi, L. SINR-based DoS Attack on Remote State Estimation: A Game-theoretic Approach, *IEEE Trans. Contr. Net. Syst.* **2016**, *4* (3), 632–642.

38. Xiao, L.; Xie, C.; Chen, T.; Dai, H. A Mobile Offloading Game Against Smart Attacks. *IEEE Access* **2016**, *4*, 2281–2291.

39. Han, G.; Xiao, L.; Poor, H. V. In *Two-dimensional Anti-jamming Communication Based on Deep Reinforcement Learning*. Proc. IEEE Int. Conf. Acoustics Speech and Signal Processing, New Orleans, LA, March 2017, pp 2087–2091.

40. Xiao, L.; Li, Y.; Huang, X.; Du, X. J. Cloud-based Malware Detection Game for Mobile Devices with Offloading. *IEEE Trans. Mobile Comput.* **2017**, *16* (10), 2742–2750.

41. Tajbakhsh, A.; Rahmati, M.; Mirzaei. A. Intrusion Detection Using Fuzzy Association Rules. *Appl. Soft. Comp.* **2009**, *9* (2), 462–469.

INDEX

Printed in the United States
by Baker & Taylor Publisher Services